TELEVISION

Electronic Pictures

These and other books are included in the
Encyclopedia of Discovery and Invention
series:

Airplanes: The Lure of Flight
Atoms: Building Blocks of Matter
Computers: Mechanical Minds
Gravity: The Universal Force
Lasers: Humanity's Magic Light
Printing Press: Ideas into Type
Radar: The Silent Detector
Television: Electronic Pictures

TELEVISION

Electronic Pictures

by LILA GANO

The ENCYCLOPEDIA of
D·I·S·C·O·V·E·R·Y
and INVENTION

P.O. Box 289011 SAN DIEGO, CA 92198-0011

DL HB FH

Copyright 1990 by Lucent Books, Inc., P.O. Box 289011,
San Diego, California, 92198-0011

Library of Congress Cataloging-in-Publication Data

Gano, Lila, 1949–
    Television: electronic pictures/by Lila Gano.
        p. cm. — (The Encyclopedia of discovery and invention)
    Includes bibliographical references and index.
    Summary: Discusses the invention, development, technology, and
future of television, its impact on various aspects of society, and
the basics of television production.
    ISBN 1–56006–202–9
    1. Television—Juvenile literature. [1. Television.]   I. Title.
II. Series.
TK6640.G36   1990
621.388—dc20   90–6470
                                                   CIP
                                                   A C

Contents

▪ ▪

■ ■

Foreword

The belief in progress has been one of the dominant forces in Western Civilization from the Scientific Revolution of the seventeenth century to the present. Embodied in the idea of progress is the conviction that each generation will be better off than the one that preceded it. Eventually, all peoples will benefit from and share in this better world. R. R. Palmer, in his *History of the Modern World*, calls this belief in progress "a kind of nonreligious faith that the conditions of human life" will continually improve as time goes on.

For over a thousand years prior to the seventeenth century, science had progressed little. Inquiry was largely discouraged, and experimentation almost nonexistent. As a result, science became regressive and discovery was ignored. Benjamin Farrington, a historian of science, characterized it this way: "Science had failed to become a real force in the life of society. Instead there had arisen a conception of science as a cycle of liberal studies for a privileged minority. Science ceased to be a means of transforming the conditions of life." In short, had this intellectual climate continued, humanity's future world would have been little more than a clone of its past.

Fortunately, these circumstances were not destined to last. By the seventeenth and eighteenth centuries, Western society was undergoing radical and favorable changes. And the changes that occurred gave rise to the notion that progress was a real force urging civilization forward. Surpluses of consumer goods were replacing substandard living conditions in most of Western Europe. Rigid class systems were giving way to social mobility. In nations like France and the United States, the lofty principles of democracy and popular sovereignty were being painted in broad, gilded strokes over the fading canvasses of monarchy and despotism.

But more significant than these social, economic, and political changes, the new age witnessed a rebirth of science. Centuries of scientific stagnation began crumbling before a spirit of scientific inquiry that spawned undreamed of technological advances. And it was the discoveries and inventions of scores of men and women that fueled these new technologies, dramatically increasing the ability of humankind to control nature—and, many believed, eventually to guide it.

It is a truism of science and technology that the results derived from observation and experimentation are not finalities. They are part of a process. Each discovery is but one piece in a continuum bridging past and present and heralding an extraordinary future. The heroic age of the Scientific Revolution was simply a start. It laid a foundation upon which succeeding generations of imaginative thinkers could build. It kindled the belief that progress is possible as long as there were gifted men and women who would respond to society's needs. When Antonie van Leeuwenhoek observed *Animalcules* (little animals) through his high-powered microscope in 1683, the discovery did not end there. Others followed who would call these "little animals" bacteria and, in time, recognize their role in the process of health and disease. Robert Koch, a German bacteriologist and winner of the Nobel prize in Physiology and Medicine, was one of these men. Koch firmly established that bacteria are responsible for causing infectious diseases. He identified, among others, the causative organisms of anthrax and tuberculosis. Alexander Fleming, another Nobel Laureate, progressed still further in the quest to understand and control bacteria. In 1928, Fleming discovered penicillin, the antibiotic wonder drug. Penicillin, and the generations of antibiotics that succeeded it, have done more to prevent premature death than any other discovery in the history of humankind. And as civilization hastens toward the twenty-first century, most agree that the conquest of van Leeuwenhoek's "little animals" will continue.

The *Encyclopedia of Discovery and Invention* examines those discoveries and inventions that have had a sweeping impact on life and thought in the modern world. Each book explores the ideas that led to the invention or discovery, and, more importantly, how the world changed and continues to change because of it. The series also highlights the people behind the achievements—the unique men and women whose singular genius and rich imagination have altered the lives of everyone. Enhanced by photographs and clearly explained technical drawings, these books are comprehensive examinations of the building blocks of human progress.

TELEVISION

Electronic Pictures

TELEVISION

Introduction

This is the age of television. Since the late 1940s, television sets have become familiar sights in homes throughout the United States. Families rely on TV as their major source of news, information, and entertainment. Some television shows also instruct viewers and can be educational.

Once this amazing invention was accepted by the public, the rapid pace of adoption of television was astounding. No other form of mass media has spread so quickly. In 1948, there were fewer than 100,000 TV sets in the United States. One year later, consumers had purchased 1 million sets. By 1959, viewers owned 50 million sets, and by 1990, more than 160 million TV sets had been placed in U.S. homes.

Why is TV so popular? Television offers affordable entertainment. Adventure, comedy, cartoons, drama, and movies are readily available on more than twenty-eight channels. Television also helps people keep up with what is happening in the world. Local news, national news, and special information programs bring important events into homes. When men landed on the moon on July 20, 1969, TV cameras transmitted this historic event to the world. "If ever there was

.. TIMELINE: TELEVISION

1   2   3   4   5   6   7   8   9

1 ■ 1835
Samuel Morse invents the telegraph.

2 ■ 1906
Radio is invented.

3 ■ 1921
The National Broadcasting Company (NBC) comes into existence.

4 ■ 1925
Charles Jenkins in the United States and John Baird in England demonstrate the first mechanical televisions.

5 ■ 1939
The first electronic television sets go on sale to the public at the World's Fair in New York City.

6 ■ 1941
NBC and CBS are granted commercial status by the Federal Communications Commission.

7 ■ 1942
The production of television sets stops so scientists and engineers can produce equipment to fight World War II.

8 ■ 1943
In an effort to control industry monopolies, the government forces NBC to give up of a major portion of its holdings. One part of the company becomes the ABC network.

9 ■ 1945
Vladimir Zworykin invents the image orthicon camera.

10 ■ 1947
President Harry Truman presents the first televised state of the union address.

11 ■ 1950–1958
Television experiences its golden age, a time marked by quality drama, excellent writing, and fine acting.

12 ■ 1951
"I Love Lucy," the first situation comedy, airs and becomes an overnight sensation. This show was the first to be filmed instead of aired live.

13 ■ 1954
The first color TV sets go on sale.

a time that television fulfilled its creators' dreams and brought the world together in peace, this was it," wrote TV historians Harry Castleman and Walter Podrazik.

Besides entertaining and educating audiences, television also convinces viewers to buy products and services. For this reason, the medium has been called "the great persuader." On commercial TV, networks such as NBC and CBS sell airtime to advertise products. Commercials have a tremendous influence on the buying habits of consumers. A clever ad can yield sales worth millions of dollars. Because TV is such a good selling medium, companies are willing to pay hundreds of thousands of dollars for a thirty-second commercial.

Not only is TV used to sell products, but politicians use it to sell themselves. The best way to introduce a candidate to voters is through political advertising on TV. With just a few seconds to get their message across, political ads try to convince voters that one candidate is far superior to another. Televised presidential debates, a campaigning tool that emerged in the 1960s, also acquaint voters with candidates. Without leaving their homes, millions of voters can see and hear the views of the candidates.

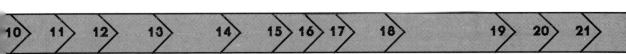

14 ■ 1960
The first televised presidential debates air. Republican Richard Nixon and Democrat John Kennedy debate campaign issues before an audience of 85 million viewers.

15 ■ 1962
a. Television is now available in sixty-five nations. Over fifty-nine million sets exist in North America.

b. An experimental satellite called *Telstar I* is launched into outer space by the National Aeronautics and Space Administration (NASA). *Telstar I* relayed the first live transatlantic TV signals from the United States to Europe.

16 ■ 1963
President John Kennedy is assassinated, and television covers the tragedy in detail.

17 ■ 1964
Live television transmissions across the Pacific Ocean become a reality using the *Relay II* satellite.

18 ■ 1969–1971
The federal government studies the issue of TV violence and aggressive behavior. Government experts find a link between them.

19 ■ 1984
The Cable Act of 1984 is passed by Congress. The law gave cable companies the right to charge subscribers whatever the market would permit.

20 ■ 1987
Fox network is born, becoming the most successful attempt to compete with the major networks, ABC, CBS, and NBC.

21 ■ 1989
A thirty-second commercial for the 1989 Super Bowl costs a record high of $700,000.

The Experimental Years

Of all the living creatures on earth, humans have developed the most sophisticated forms of communication. The search for better ways to communicate began thousands of years ago and will continue well into the future. With the discovery of electricity came inventions such as the telephone, telegraph, radio, and television. Modern technology allows people to communicate ideas, sounds, and pictures all over the world, but for centuries progress was very slow.

As early as 1267, a scientist named Roger Bacon wrote about sending signals through the air using electricity. But in those times, superstition and lack of scientific understanding caused people to fear Bacon's theory. His ideas were so frightening and revolutionary that he was imprisoned for "dealing in black magic."

During the 1700s and the 1800s, people became more interested in science, and scientists experimented with the fundamentals of electricity and magnetism. With each experiment came more understanding, sometimes obtained by accident. By the nineteenth century, the stage was set for a major breakthrough in the form of an invention called the telegraph. In 1835, Samuel Morse, a college professor at New York University, demonstrated an electromagnetic device that transmitted messages by wire. Using electrical signals that corresponded to letters of the alphabet, Morse used the telegraph to

Samuel Morse began the communications revolution in 1835 by inventing the telegraph and Morse code. Morse's invention proved that sending messages by electrical impulses was a scientific reality—not magic.

tap out messages that traveled almost instantaneously over wires strung from Washington, D.C., to Baltimore, Maryland. His system of dots (short taps) and dashes (long taps) became known as Morse code.

As the telegraph spread across the country, it helped bring Americans closer together and opened up the Western frontier. News traveled across the continent in minutes instead of days, and business and personal communications became easier. The telegraph also helped pave the way for both radio and television by proving that communication via electrical impulses was a scientific reality, not "black magic."

The Invention of Radio

By 1906, inventors had developed a method of transmitting voices through the air without wires. They found that sound traveled on radio waves, a form of electromagnetic waves. This invention was called radio. At first, radio transmission was just a hobby for amateur radio operators all over the country. Later, it proved to be very important in helping win World War I. During the war, radio was an excellent way to share information about enemy sightings, coordinate sea rescues, and keep in contact. The new device was praised for its role in aiding war efforts. Hardly anyone expected this practical invention to entertain and educate the country for the next four decades.

After the war, radio manufacturers looked for nonmilitary ways to use radio. Experimental broadcasts on a small scale convinced them that the public was indeed interested in receiving the sounds of radio into their homes. The first radio station featuring regular broadcasts went on the air in the fall of 1920. That night, radio grabbed the nation's attention by covering the presidential election results. A live band and selected vocalists filled the night with music between election announcements. The first boxing match was broadcast a year later, prompting thousands of people to buy radios for that event alone. Broadcasts of major sporting events, such as the

Graham McNamee (left) and Philips Carlin broadcast the opening game of the World Series at Yankee Stadium over the radio on October 4, 1928. Broadcasts of major sporting events created a surge of public interest in radio, and later, television.

World Series, seemed to create a surge of interest in radio. This trend also helped popularize television years later.

In those days, radio manufacturers like General Electric (GE) and the Radio Corporation of America (RCA) entered the broadcasting business to spur the sale of radios. The idea worked well. "The era of radio expansion had begun, creating one of the most extraordinary new product demands in the history of the United States," wrote Irving Settel in his book, *A Pictorial History of Radio*. Demand was so intense that manufacturers could not produce enough sets to satisfy buyers.

Broadcasting stations, which transmit radio programs to homes, sprang up all over the country. In 1926, nineteen stations came together under one umbrella company known as the National Broadcasting Company (NBC). A few years later, forty-seven stations joined to form the Columbia Broadcasting Station (CBS), a rival network.

Still in their infancy, radio broadcasters needed a way to make money to support themselves. The stations hit upon the idea of selling airtime to companies that wanted to advertise their products. The radio stations then used the profits to finance more creative shows and to pay salaries for performers. Ultimately, these advertisements lured bigger audiences to the radio programs. Advertisements on radio became known as commercials.

The Impact of Radio

Radio quickly worked its way into millions of homes, changing the fabric of family life and American society. The voice of radio became the voice of the outside world. Families everywhere learned about current events soon after they occurred, no matter where they took place. Radio informed listeners about Charles Lindbergh's historic flight across the Atlantic Ocean in 1927. Radio covered political events and reported events happening overseas. Politicians both in and out of office learned to use radio to reach vast numbers of voters.

In 1929, radio helped bring incredible news to a nation that had enjoyed a decade of prosperity. The country suddenly found itself in a state of economic disaster, primarily because of losses in the stock market. Banks closed, companies went out of business, and countless

Aviator Charles Lindbergh flew the Spirit of St. Louis *from New York to Paris in 1927, the first solo flight across the Atlantic. Radio made it possible to hear about historic events like this almost as soon as they occurred.*

people lost their jobs and homes. The years following the economic plunge were known as the Great Depression.

For many people, living in poverty or close to it, radio proved to be a beacon of hope. President Franklin Delano Roosevelt comforted a desperate nation through a series of talks known as "fireside chats" transmitted via radio. He explained what was happening in the country and what he was doing about the financial problems. "This was the first time that the American people had been spoken to so simply and directly by their president," wrote Settel. Radio was also a free source of entertainment available to millions of poor people. Families sold their cars and furniture to buy food, but few would part with their radios. So great was the public's re-

Radio allowed national leaders like President Franklin D. Roosevelt to communicate directly with the public. President Roosevelt regularly used radio broadcasts to inform the people of his policies and to encourage them during the Great Depression.

Actor/writer Orson Welles panicked the whole nation on Halloween, 1938 by broadcasting a dramatic reading of H. G. Wells' story War of the Worlds. *Many listeners believed they were hearing a newscast about a real invasion from Mars!*

liance on radio during the depression years that the radio industry boomed while other industries struggled.

By 1930, over nine million radios had been sold across the nation. Radio sales reached hidden pockets of people, from the mountaintops of Appalachia in Virginia to isolated ranches in Wyoming. Through radio, people shared the same national and international news, laughed at the same comedies, and cried at the same dramas. In this way, radio helped shape a common culture and unify the nation in times of crisis.

But in one famous broadcast, radio created its own crisis. On Halloween night in 1938, Orson Welles, a young actor and writer, presented on the radio a well-known science fiction story about an invasion from Mars. The story

unfolded in such a realistic fashion that many listeners actually believed it was a genuine news broadcast. People across the nation panicked. Some families barricaded themselves behind closed doors while thousands more fled their homes, trying to escape from the Martians. Calm was restored the next day amid apologies from the radio station. Nonetheless, this event demonstrated the almost hypnotic hold that radio had over listeners and just how trusted the medium had become.

Radio was a "theater of the mind" and challenged listeners to imagine the pictures to accompany the sounds. Audiences were entertained royally night after night, day after day. They laughed at the zany comedy shows, such as "The Fred Allen Show" and

Actors perform a comic skit for the popular "Fred Allen Radio Show" in the 1940s. The programs were viewed live by a small studio-audience, but the majority of people listened to them over the radio. Radio programs quickly became an important part of American family life.

"Baby Snooks." During the daytime, intense dramatic sagas entertained housewives while they took care of domestic duties. These shows were commonly called soap operas because they were usually sponsored by soap manufacturers. Musicals, cooking demonstrations, and talk shows featuring colorful hosts enthralled audiences for decades, much the way television would later. Families all over America gathered together at the end of the day to enjoy favorite radio programs and personalities. For many homes, family life revolved around the radio, especially in the evenings.

Motion Pictures

While radio entertained families inside the home, outside the home, Americans embraced another form of enter-tainment—motion pictures. Whereas radio offered sound with no pictures, the first motion pictures, known as "flickers," offered pictures with no sound. To help the audience follow the story, dialogue and descriptions about the scene were written and displayed on the bottom of the screen. Even without sound, audiences adored the comedy routines of Charlie Chaplin and the romantic adventures of Rudolph Valentino. Admission to a silent movie cost about a nickel, a small price to pay for lively entertainment.

In 1927, motion picture technology made a giant leap forward by adding sound to the pictures. With the introduction of "talkies," motion pictures became an even bigger hit with the public. During the depression, movies as well as radio helped Americans forget their terrible problems for a short time.

The public was so hungry for movies that about five thousand were made between 1930 and 1940. Going to the movies had become a favorite American pastime.

Bringing Motion Pictures into the Home

Much money had been made from the radio and motion picture industries. With the great success of these two forms of entertainment, businesspeople and scientists looked to the future. What if scientists could invent a small device that could combine pictures and sound—a product that would fit into living rooms all over the country? It was not long before the dream became a reality. It was called television.

Tele comes from the ancient Greek language and means "at a distance." The word *television* means to see, or view, over a distance. The concept of television first appeared in a French publication in 1900, but the invention was not perfected until many years later.

The progress in the research that eventually brought us the bright, clear pictures on modern TV sets began in the early 1920s. The birth of television technology occurred in the laboratories of several countries, including England, Germany, and the United States. Inventors wanted to transmit quality pictures and sound through the air into a receiving unit, the TV set. But design problems took a long time to overcome, and TV technology advanced slowly. Inventors realized that they needed more than their ingenuity to perfect TV. They also needed a lot of money to pay for equipment and experiments.

The development of TV technology relied heavily on money from big businesses. The companies most interested in TV research were the large radio manufacturers and broadcast networks. RCA, GE, and Westinghouse wanted to build and sell televisions as well as radios. Radio broadcast companies, like NBC and CBS, not only produced and transmitted popular radio shows to the public but also conducted their own TV research. Profits from the sale of commercials on radio were reinvested into television research. With the fabulous success of radio, everyone believed television to be the next great moneymaker in the field of home entertainment and communication.

Not all TV research was conducted by big radio companies. Some inventors worked independently in their own labs. Most still needed money to pay for experiments. These independent inventors teamed up with investors who paid for experiments. Investors were usually businesspeople, such as bankers, who had extra money that they wanted to invest in a promising project. In return for financial backing, investors were promised part of the profits from sales of the invention, if it were successful.

Two of the earliest inventors who moved TV technology a major step forward were independents. Amazingly, both performed similar experiments during the same year, 1925. Though these inventors used a similar process, they were thousands of miles apart. One worked in the United States, and the other worked in London, England.

In the United States, Charles F. Jenkins relayed TV signals from an old Navy radio station to his laboratory in Washington, D.C. TV signals traveled a distance of five miles between the two

Scientist and television authority D. E. Replogle gives the first public demonstration of talking moving pictures transmitted by radio. His audience gathers around the receiver, called a Jenkins Radiovisor, in a 1920s Newark, New Jersey piano studio.

sites. With an excited audience of military and civilian dignitaries in his lab, Jenkins conducted his magical demonstration. On a screen, he produced im-

Dr. C. F. Jenkins, a pioneer in television and motion picture technology, tunes in his latest invention: a television that can receive words as well as visual images. The viewer watched the screen at the top, while hearing the sound from the speaker at the bottom.

ages of a windmill's rotating vanes and footage from an old Hollywood movie.

In London, inventor John L. Baird, a Scotsman, produced images of a puppet's head on a screen. Though the pictures were dim and fuzzy, Baird continued his experiments. By 1929, he began broadcasting shows in London. Baird's broadcasts were more experimental than entertaining. No Saturday morning cartoons or cop shows existed then. Few people even owned or wanted to own television sets at that time because the picture quality was so poor.

Mechanical Television

The method that Baird and Jenkins used was known as mechanical television. Anything mechanical relies on the use of machinery to operate it or to make it move. Mechanical television was comparable to the illusion of movement created when a group of still pictures is flipped so fast that they appear to move. Two spinning disks provided the movement in mechanical TV.

Scottish inventor John L. Baird was developing television technology in England at the same time that C. F. Jenkins was working on it in America. Baird even experimented with color television. His work proved that color TV was possible.

The spinning disks resembled paper plates but had holes cut out in a circular pattern from the center to the edge. As the first disk spun in front of the subject, light passed through each disk hole. Each hole "captured" a small bit of light and part of the subject's image. The light then traveled through a photoelectric cell that changed it into electrical signals. With the image now in the form of an electrical signal, it had to be transferred back into something the human eye could see. This was done by sending the electrical signals through another spinning disk that reconstructed the original image on a screen. These early images were black-and-white, but experiments with color were soon to follow.

By adding a color filter to the process, Baird transmitted the first color pictures in London in 1929. The first color image he transmitted was a man sticking out his tongue. Later on, a red and blue scarf took center stage to show off its colors. Though these pictures were primitive compared with the clarity on modern color sets, Baird proved color TV was possible.

With the success of Baird and Jenkins's experiments, sales of mechanical black-and-white television very slowly spread in England and in America. (Color TV would take a backseat for the next two decades.) But mechanical systems had several drawbacks. First, the equipment was large and bulky. Few peo-

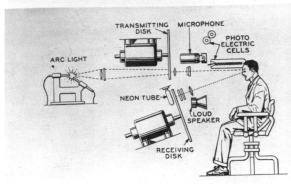

Early videophone equipment was large and bulky. The diagram shows how the equipment works. Using a combination of telephone and television technology, this equipment enabled the user to both see and hear the person at the other end of the transmission.

MECHANICAL TELEVISION

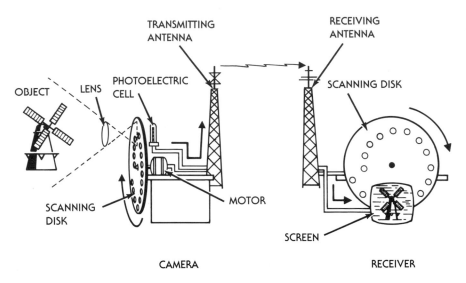

The mechanical television system developed by John Baird in 1929 used two rotating scanning disks, one in the camera and another in the receiver. Each disk had a series of sixteen holes on its surface, arranged in a spiral.

The objects in front of the camera reflected different amounts of light through the camera lens onto the first scanning disk. As the disk spun, the holes broke the reflected light into a sequence of light pulses of varying intensity. A photoelectric cell located behind the disk converted the varying pulses of light into a sequence of electrical signals that could be transmitted to a television receiver.

At the receiver, the process was reversed. The sequence of electric signals caused a lamp to flash with varying brightness. The light from these flashes passed through a second scanning disk, which was synchronized exactly with the first disk. The spinning disk transformed the flashing light into light and dark spots that appeared on a screen. When viewed from a distance, these spots made up an image resembling the one viewed through the camera.

ple wanted a huge, awkward machine in their living room. Second, the picture quality was poor. Third, performers were prone to faint under the bright lights required by the mechanical approach. Scientists and inventors looked for a better design to overcome some of these drawbacks. The path to improved TV technology led early researchers into the science of electronics. Ultimately, they abandoned mechanical TV.

Electronic Television

Electronics is the science of electrons, how they behave, and the ways their activities can be harnessed for inventions. Electrons are the smallest unit of electricity in existence, and they revolve around the center of atoms. When electrons are placed in a container without air, a vacuum, they behave in unique ways. Understanding the behavior pat-

WHAT IS AN ELECTRON?

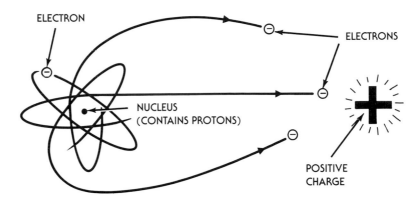

ELECTRON

ELECTRONS

NUCLEUS
(CONTAINS PROTONS)

POSITIVE
CHARGE

Electrons are negatively charged particles that orbit around the nucleus of an atom. Some electrons in the outermost orbit of an atom can be drawn away from the atom by a strong positive charge. This can create a flow of electrons. If the electrons flow through an airtight tube, they make up an electron beam, which is a component of electronic television. Electrons can also be conducted through wires or other solid materials. This is called an electric current.

terns of electrons led scientists to the invention of electronic television.

No one person can take full credit for inventing electronic television, which is what we use today. The credit is primarily shared by two men, Vladimir Zworykin, a Russian, and an American named Philo Farnsworth. Both were inspired by the work of Professor Boris Rosing.

Zworykin was a graduate student working under Professor Rosing in Russia. As early as 1907, Rosing looked for a way to electronically send visual images. He finally succeeded in transmitting crude pictures in his laboratory. When Zworykin came to the United States around 1920, he began his own experiments based on the teachings of Rosing. Zworykin's work began in the labs of Westinghouse. Later, his experiments were moved to RCA where he received financial support.

The Iconoscope and Kinescope

RCA's president, David Sarnoff, was determined to bring television into the lives of Americans at almost any cost. Under Sarnoff's guidance, RCA pumped millions of dollars into Zworykin's TV research. By 1928, the experiments showed major progress. Zworykin developed two electronic devices that helped lay the foundation for modern television. These were the iconoscope, or the TV camera tube, and the kinescope, or the picture tube inside the set. Both devices operated on the principle of controlling electrons in a vacuum tube.

The camera tube was a vacuum tube with a small screen inside of it. With the help of a camera lens outside of the tube, an image was projected onto one

ICONOSCOPE AND KINESCOPE

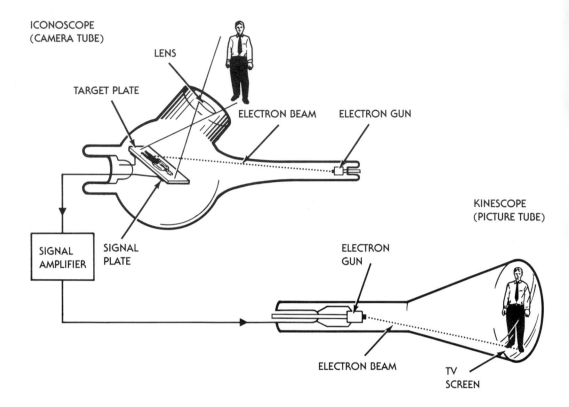

The iconoscope, or camera tube, and kinescope, or picture tube, were perfected by Vladimir Zworykin in the 1920s. They were the basis for modern television transmission.

In Zworykin's iconoscope, reflected light from an image in front of the camera was projected through the lens onto a target plate. The target plate had a photosensitive coating, which was altered when struck by light. On the back side of the target plate was a signal plate.

An electron gun in the iconoscope shot an electron beam that scanned the target plate. The number of electrons that passed through to the signal plate varied exactly the same as the photosensitive coating on the target plate had been varied by the light. The signal plate translated the varying number of electrons into an electrical signal, which was sent to a signal amplifier.

The signal amplifier sent electrical signals to the kinescope, or picture tube. Here the signal triggered an electron gun to emit a varying number of electrons in a beam. This electron beam scanned the front of the picture tube, which had a fluorescent coating that glowed wherever it was struck by electrons. The more electrons that hit any one spot, the brighter that spot would glow. The front of the picture tube was also the screen of the television, and its variously glowing spots reproduced the camera image.

side of the screen. On the other side of the screen inside the tube, beams of electrons passed over every part of the image, moving from side to side in a sequence of horizontal lines. This process was called scanning, or reading, and is still used in modern TV systems. As beams of electrons scanned the original image in the camera tube, they created electrical signals that were transmitted to a glass picture tube known as the kinescope. Zworykin's picture tube was shaped like a large funnel with a long neck. At the base of the thin neck, electrical signals from the iconoscope entered and rapidly traveled toward their target—the picture screen at the opposite end. As the electrons swept across the screen, they reassembled the picture in its original pattern. The more times that the electrons traveled back and forth, the more lines they created, thus building a clearer picture. In mechanical TV, the images were composed of thirty to sixty lines, and the pictures were fuzzy. Electronic scanning created a clearer picture by doubling the number of lines.

The Image Dissector

In Idaho, a young farm boy named Philo Farnsworth dreamed of building an electronic television system. Though thousands of miles from Russia, Farnsworth had read about Rosing's work in a magazine article and knew he could improve upon the design. While plowing the fields on the family farm one day, Farnsworth discovered the same principle of electronic scanning that Zworykin was using in his research. "He turned at this little high spot [in the field] to see if the rows were straight,

Philo Farnsworth (center) demonstrates the workings of an early television camera. An Idaho farmboy, Farnsworth had diagrammed a workable TV system at the age of fifteen.

and it just hit him like a thunderbolt, 'I can scan a picture that way, by taking the dots, the electrons, back and forth as you would read a page,'" his wife, Elma, said later. By the time Farnsworth was fifteen years old, he had diagramed a workable, electronic TV system.

By his twenty-first birthday, Farnsworth's ideas captured the interests of several investors. They agreed to help pay for his independent research. The investors put up a total of $25,000 for the first experiments, a large sum of money in 1925. By the time Farnsworth had achieved a quality picture several years later, investors had contributed over $1 million toward the research.

Farnsworth's electronic TV system was based on the same principles as Zworykin's iconoscope and kinescope, with a major exception. Farnsworth developed a better device for scanning the picture to be transmitted. He called this device the "image dissector." The dissector broke the image down into 150 lines and scanned it thirty times per second, producing a clearer picture than what the iconoscope and kinescope produced. Farnsworth's work improved the quality of the TV picture and helped bring television out of the lab and into the marketplace.

Tough Times for Television

By 1939, electronic TV was ready to be unveiled before the public. TV sets made their debut, and for the very first time consumers could buy electronic TV sets for home use. The sets were manufactured by RCA, which had obtained the rights to use Farnsworth's version of electronic TV. To ensure that electronic TVs received a great deal of attention, David Sarnoff, RCA's power-

RCA president David Sarnoff used the setting of the 1939 New York World's Fair to "advertise" RCA's new product: electronic television sets. Despite the attention he got by broadcasting President Roosevelt's opening day speech, the public did not buy televisions.

ful president, introduced them at the World's Fair in New York City that same year. Sarnoff also made arrangements to televise the World's Fair with President Franklin Roosevelt presenting an opening day speech. He became the first president to appear on television. But despite all the publicity and excitement, TV sales were disappointing.

The public was not buying the product that Sarnoff was pushing. Electronic TV still had some undesirable features. The sets were bulky, though less so than mechanical sets. Of the twenty-five models on display at the fair, several included a complex mirror arrangement. Viewers were supposed to watch the screen image reflected off the mirror because it was feared that looking directly into the screen might damage the eyes. The price of the sets ranged from two hundred to one thousand dollars. Few consumers stormed the World's Fair to buy these electronic wonders. By the end of the year, fewer than one thousand sets had been sold. What was the TV industry doing wrong?

Many people believed that the lack of public response was because there were so few TV stations to broadcast, or transmit, shows. This meant that few people could actually receive broadcasts. In 1939, less than ten TV broadcasting stations were in operation throughout the entire country.

Broadcasting means to send out TV signals in all directions for televisions everywhere to pick up. TV signals are pictures and sound that have been changed into electrical impulses. This process occurs at a TV broadcasting station and is the first step in getting TV shows into viewers' TV sets. TV stations control what programs viewers see and when they see them. After the signals

AN ELECTROMAGNETIC WAVE

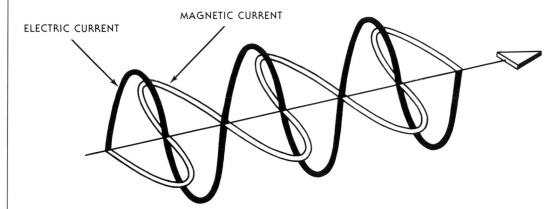

ELECTRIC CURRENT

MAGNETIC CURRENT

A radio wave is an electromagnetic wave. This is a form of energy containing both an electric current and a magnetic current. Both currents travel in the pattern of a wave at the speed of light. In each electromagnetic wave, the electric current and magnetic current have exactly the same wave pattern, or frequency. These waves carry sound and picture signals from radio and TV stations to radio and TV sets.

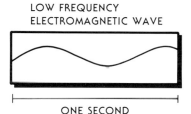

LOW FREQUENCY
ELECTROMAGNETIC WAVE

ONE SECOND

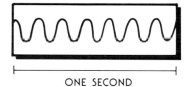

HIGH FREQUENCY
ELECTROMAGNETIC WAVE

ONE SECOND

Electromagnetic waves travel in all different frequencies. A low- frequency wave has only a few peaks per second, while a high-frequency wave has thousands of peaks per second. The higher the wave's frequency, the more information it can carry.

---

leave the broadcasting station, their next stop is a transmitting station.

Complicated equipment at the transmitting station combines the sound and picture signals with radio waves. These waves are called carrier waves because the TV signals "hitchhike" a ride on them to the viewer's home. Carrier waves travel at the incredibly fast speed of 186,280 miles per second.

They travel in all directions from the tops of tall towers called transmitter masts. Because the waves tend to bounce off of obstacles in their paths, transmitters are usually placed on high buildings or hills to avoid this. If a carrier wave is blocked, poor TV reception may result.

Carrier waves are limited in other ways. First, they tend to weaken after traveling long distances. Second, they travel in a straight line and cannot follow the rounded contour of the earth's surface. But technology has overcome these limitations by simply locating transmitter masts close together. As carrier waves travel from transmitter to transmitter, special equipment called an amplifier helps strengthen the waves.

After leaving the transmitter, the carrier waves reach the antennae on the viewers' rooftops. Antennae are made of metal that has been twisted into various shapes. The shape and the electromagnetic characteristics of antennae attract the carrier waves. When the viewers turn on TV sets, the waves picked up by the antennae enter the sets.

When television was introduced, the programs that were broadcast were not popular with the general public. Boxing matches and New York plays appealed to specific groups but not to the masses. Shows were different from the ones of today. Also, the first TV shows were not elaborate, and there were no weekly comedies and dramas. Most of the programming served the New York and Philadelphia areas and ignored the rest of the nation. Who could blame the public for clinging to their radios for quality entertainment with well-written scripts and characters they had grown to love? TV had very little to offer.

Television Bombs During World War II

Americans also had more important things to think about than new forms of entertainment. World War II had broken out in Europe and threatened to draw the United States into battle. Germany wanted to take over Europe, and Japan threatened the islands of the South Pacific. In 1941, Japan bombed U.S. Navy ships at Pearl Harbor, Hawaii. President Roosevelt and Congress reacted by declaring war on Japan and then on Germany and Italy four days later.

CBS leaped on the story of the Pearl Harbor attack. In a TV broadcast that lasted nine hours, newsman Dick Hubbel covered the dramatic events surrounding the surprise bombing. Television could have played a very important role in reporting war developments to the home front, but just the opposite happened. When the nation entered World War II, TV factories were ordered to make equipment to support the fighting. TV sets were considered a luxury item not vital to the war effort. The nation had to devote all of its resources to building guns, ships, and other materials to confront the enemy.

Not only did the production of TV sets stop but most television broadcasting stations shut down. The technically skilled men and women who ran the stations were needed for the war effort. The few stations that remained in operation cut back their broadcasting schedules. In New York, the stations reduced broadcasting from four to just one night per week. With only a few thousand sets in existence, most Americans relied on their radios and newspapers to keep them informed of world events. Television had missed the opportunity to demonstrate its potential as a great communication device.

Business was definitely slow in the TV industry in the early 1940s, but one important development did occur. A new broadcast network was born—the American Broadcasting Company (ABC).

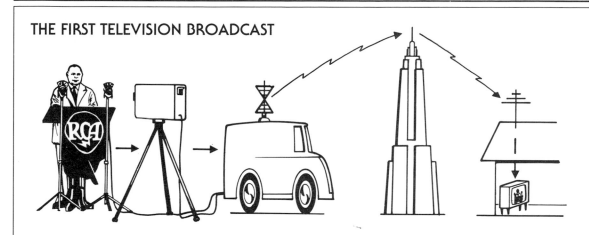

THE FIRST TELEVISION BROADCAST

The first live television broadcast in the United States took place at the New York World's Fair in 1939. An iconoscope television camera transmitted images of David Sarnoff, president of RCA, from the World's Fair to about five hundred television receivers placed throughout the city.

The transmitting station was located in the back of an RCA van. It converted the sound and picture signals into radio waves and broadcast them so they would be received by a radio antenna on top of the Empire State Building. From there, the waves were rebroadcast so that they could be received by television sets throughout the city.

Up until this time, CBS and NBC had a virtual monopoly over the airwaves. Their only serious competition was the DuMont network, a small company that ran into serious financial problems and finally folded in the mid-fifties. ABC came into existence because the government wanted there to be more competition in television broadcasting. As the industry's regulator, the Federal Communications Commission (FCC) forced NBC to split up and form a new, separate network, which was ABC. In the end, viewers would have more programming choices because of the government's actions.

As World War II drew to a close, the television industry slowly revived. CBS adapted successful radio shows to television. One early show that had its roots in radio became a hit TV series. The show was called "Missus Goes a' Shopping" and presented silly antics that made people laugh. An overweight truck driver who tried to put on a girdle was one of the stunts that won this show rave reviews. The visual appeal of this show helped attract attention to TV. By the end of the war in 1945, television was poised on the edge of a new era of acceptance.

Television Comes Alive

In 1945 television reawakened to a new era of opportunity. TV stations reopened, now that the war was won, and new ones were built all over the country. Producers experimented with different formats for shows, like cooking demonstrations and quiz games. To add to the momentum, an exciting technical advance renewed the public's interest in television and won over critics. A new camera called the image orthicon unlocked TV's potential as a visual medium. No significant improvements in camera equipment had occurred since the late 1920s. Designed by Vladimir Zworykin in 1945, the new camera solved two significant problems.

With the older cameras, extremely bright and hot lights were needed to shoot a scene. These cameras were not very sensitive to light and therefore required a great deal of it to "see" the image. Performers working under the lights became uncomfortable quickly. The heat sometimes caused them to faint and melted their makeup. The image orthicon made TV performing more comfortable for actors and actresses because less light was needed. The new camera was one hundred times more sensitive to light than previous models and also produced a much clearer picture.

The image orthicon camera also solved another problem. Early TV cameras could not film events that took place over a large area, such as a foot-

Vladimir Zworykin inspects his 1945 invention, the image orthicon camera. The orthicon was one hundred times more sensitive to light than previous TV cameras. It not only provided a clearer picture, but greatly reduced the need to use so many hot, bright lights during filming.

ball game or a play on a large stage. They had a limited depth of field. This meant that images close to the camera were in focus and those that were farther away appeared somewhat blurred. The image orthicon allowed for greater depth of field and made it possible to televise sporting extravaganzas, like fights from large sports arenas.

The improved camera made a dramatic impact on TV audiences, particularly sports enthusiasts. The first World Series baseball games were televised in 1947 using the image orthicon. This presentation attracted

IMAGE ORTHICON CAMERA

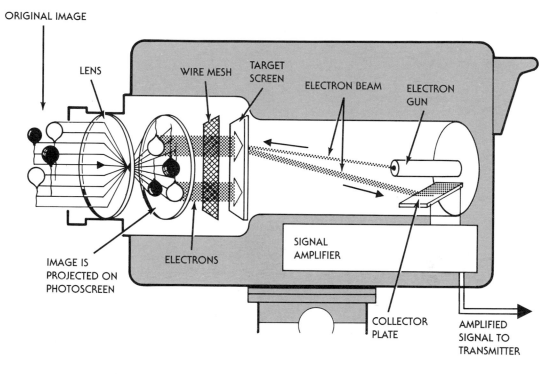

The image orthicon camera, developed by Zworykin in 1945, significantly improved the focus and sharpness of the televised image. Just behind the lens in the image orthicon camera is a small screen called a photoscreen. When light strikes the photoscreen, it gives off electrons. The more light that strikes it, the more electrons it gives off.

The electrons pass through a wire mesh, which arranges them into rows and columns. An electron gun at the back of the camera shoots a scanning beam that reads the electron pattern on the target screen. This beam is reflected onto a collector plate, which transfers the electron pattern into an electric signal, which it sends to a signal amplifier. The amplified signal is then sent to a transmitter and broadcast to television sets, which reconstruct the original image.

about 3.9 million viewers. Since home TV sets were still scarce, how did so many people get to watch the Series? They went to their neighborhood bars. Bars not only served beer and pretzels but also offered endless hours of free television entertainment. About 3.5 million eager sports fans flocked to their local bars to watch the Series. No doubt, this event and the clarity of the picture from the image orthicon convinced many consumers that they needed a TV set for their home. The television "epidemic" spread rapidly after the World Series. Families lucky enough to buy the first set in their neighborhood became instant celebrities. Friends and neighbors stopped by

(Top) In 1950, people at a downtown diner in Kansas City, Missouri seem to forget their meals as they view a World Series baseball game broadcast on television for the first time.
(Bottom) One of two TV cameras captures the action of the first football game ever to be telecast. The year was 1939. The other camera was up in the press box and was used to get whole-field views. The sideline camera pictured here was used for close-ups.

to watch the magic of television and were captivated by its spell.

The production levels of TV sets skyrocketed from 178,000 in 1947 to 975,000 in 1948. TV manufacturers could hardly keep up with the public's demand for their product. Over 4.4 million sets were in use as television ushered in the next decade. As TV found its way into homes across the nation, it slowly eroded radio's position as the favorite form of home entertainment. The motion picture industry also suffered. As programming improved, viewers spent more time in front of the TV and less time at theaters. Why go to the movies when TV could entertain for hours and for less money? Clearly, TV was changing the habits of Americans.

Color TV Makes Its Debut

Businesspeople in the television industry were eager to develop color TV. They believed that color would make TV sets more appealing and increase sales. Once again, the race to perfect a workable color TV started and ended in the laboratory. CBS and NBC were in serious competition to get their systems approved by the FCC.

At the CBS laboratory, Dr. Peter Goldmark was in charge. In 1940, he unveiled a workable color system. His system used some of the principles of both mechanical and electronic television. Goldmark's approach was mechanical because it relied on a spinning color drum, or wheel, inside the camera. A small motor caused the drum to spin. As the drum rotated, blue, red, and green filters on a wheel moved in front of a pickup tube where the image was formed and scanned by beams of electrons, a key feature of electronic TV. After the image was scanned and turned into electrical impulses, the impulses left the camera through a cable and traveled to a transmitter, finally ar-

An American family in 1948 enjoys a television program in the comfort of their living room. As TV technology and programming improved, people began to watch television instead of going out to the movie theater. The popularity of TV watching increased even more with the development of color television.

riving at the picture tube inside the TV set. The receiving TV set also contained a spinning drum with a color wheel that was in synchronization with the one in the camera. The spinning drum in the TV set produced high-quality color images in Goldmark's early version of color TV.

Success in the laboratory met with resistance from the FCC. The color wheel would be too big and impractical for home TV sets. The system had another fatal flaw. When color transmissions went out over the air using Goldmark's invention, black-and-white sets received no picture, just static. If Goldmark's system had been accepted by the FCC and all broadcasts made in color, black-and-white sets would have become useless. Nonetheless, CBS pushed for approval of its system as the standard. After much debate, the FCC told CBS that its system was not ready for commercial use. What was needed was a fully electronic color system that allowed black-and-white sets to pick up transmissions, even though they would not be in color.

By 1953, in a surprise move, RCA leaped ahead of CBS in the color race. RCA scientists invented a fully electronic system that was compatible with black-

and-white sets. This meant that owners of black-and-white sets could receive clear black-and-white pictures (not static), even though the original broadcast was in color. David Sarnoff convinced the FCC to approve RCA's system and scored another victory for his company. Within a year, the first color TV sets, made by RCA, appeared on the shelves. The electronics of these early sets was very similar to modern sets of today.

Inside the Color Set

As the viewer tunes in to a favorite cartoon or soap opera, antenna cables transport the signals from the broadcast station to the TV set. Inside the set, the electrical signals are separated into visual (picture) and audio (sound) signals. An amplifier strengthens the audio signals that come out as voices and music through the set's speaker. The visual signals travel via a different path into the picture tube.

At the narrow end of the picture tube are devices called electron guns. After power is supplied in the form of electricity, the guns release electrons. Most color TV sets have three guns to

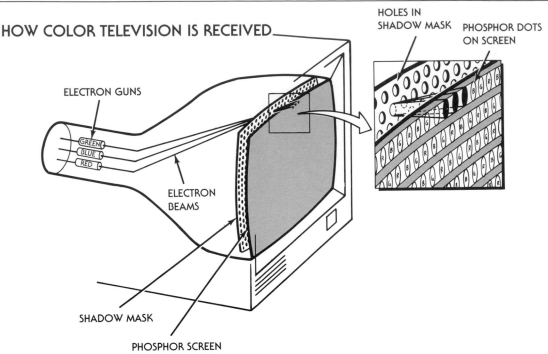

HOW COLOR TELEVISION IS RECEIVED

ELECTRON GUNS

GREEN
BLUE
RED

ELECTRON BEAMS

SHADOW MASK

PHOSPHOR SCREEN

HOLES IN SHADOW MASK

PHOSPHOR DOTS ON SCREEN

A color television picture tube has three electron guns, one for each of the three primary colors—green, blue, and red. The TV screen is covered with over 300,000 colored phosphor dots that glow when struck by electrons. The dots are arranged in groups of three, with a green, blue, and red dot in each group. When the electron guns receive a signal, they shoot beams at their respective dots. Before striking the screen, the beams pass through a metal shadow mask that keeps each electron beam in line with its own color dots and away from dots of other colors. The variations of thousands of glowing dots of green, blue, and red appear to the eye as images of all different colors.

control electrical signals for three different colors—red, blue, and green. (Black-and-white TVs have one gun.) Though they are not weapons in the usual sense, the guns "shoot" beams of electrons toward the screen at the other end of the picture tube. To help guide the electrons toward the viewing screen, a metal plate called a shadow mask lies behind the viewing screen. The shadow mask is drilled with thousands of tiny holes. These holes help focus the electron beams from each gun on the back of the TV screen. (Black-and-white sets do not need a shadow mask because they have only one electron beam.)

The back of the screen is called the faceplate. It is covered with millions of tiny colored dots or strips of phosphorus, a substance that gives off light when excited by radiation. The faceplate is the source of color. In color sets, each dot or strip is made up of three colors—red, blue, and green. The red part glows when hit by electrons from the red gun, the blue parts

TELEVISION TRICKS THE EYE

An image on a television screen is actually composed of hundreds of lines. But to a viewer stationed at least a few feet away from the set, the lines appear as a solid image.

glow when electrons from the blue gun hit it, and so forth. As the colors glow in different combinations, the lenses of the viewer's eyes are tricked into seeing a rainbow of many different colors, including purple, yellow, and flesh tones.

The electrons move in a strict path. They travel back and forth in a horizontal pattern, skipping a line and scanning alternate lines as they move. In U.S. television sets, they make 525 trips and create the same number of lines to build the picture. Electrons move too fast for the viewer to notice any movement. If the viewer sits several feet from the screen, the lines blend together to create a distinct image, but up close, the individual lines can be seen.

TV Broadcast Stations

Increased TV set production was accompanied by an expansion of television broadcast stations. By 1951, NBC had built a network, or chain, of television broadcast stations from coast-to-coast. With about sixty-one broadcast stations, NBC could relay programs to most of the major cities. The coast-to-coast connection meant that viewers in Los Angeles, for example, could see the same shows at the same time as viewers in Boston. For West Coast viewers, this meant that a show actually appeared three hours earlier because of the time zone difference.

Television studio 8-G was officially opened by NBC at Rockefeller Center in New York City on April 22, 1948. It is shown set up for four different broadcasts.

The Early TV Shows

Television programming in the late 1940s borrowed ideas from motion pictures. One of the first westerns to appear on TV was "Hopalong Cassidy." Before galloping onto the TV screen, this cowboy hero, played by William Boyd, appeared in sixty-six movies over almost two decades. When television became popular, the movie versions of "Hopalong Cassidy" were edited into shows for television. In 1951, William Boyd began producing Cassidy shows made especially for TV.

The "Lone Ranger" was another TV western that spanned entertainment mediums. This popular children's show featured a mysterious masked man and his partner, Tonto, a Native American. They roamed the old West helping others out of bad situations and ensuring that justice prevailed. The "Lone Ranger" was first created for radio audiences in 1933. Later, a series of fifteen episodes

was made for movie fans. In 1949, the Lone Ranger and Tonto stormed television and stayed for six years. But westerns were not the only kind of entertainment that excited TV audiences.

Bright comedians combined their talents with creative writers and made comedy shows a favorite. Audiences went wild over the visual antics and pranks of an amazing entertainer named Milton Berle. Berle had been on radio but had not been very successful. His funny costumes and slapstick jokes, however, were perfect for television. His show was called "The Texaco Star Theater" and was on NBC for eight years.

Milton Berle was the essence of the best of television comedy. For that reason, he earned the title "Mr. Television." Berle's show was so funny that he was accused of "emptying city streets on Tuesday nights." People everywhere reorganized their schedules to stay home and watch Milton Berle. "A

Comedian Milton Berle's genius for visual comedy made him the first TV superstar and earned him the title "Mr. Television." His show, "The Texaco Star Theater," ran for eight years on NBC in the 1950s. Comedy variety shows were one of the early successes of TV programming, along with westerns and children's programs.

This all-black cast of the televised version of "Amos 'n' Andy" replaced the original radio program's white cast. The hugely popular radio program ran for twenty-five years, but the TV version aired for only two years because of the controversy it raised. Leaders of America's black community claimed that the show reinforced negative stereotypes of black people.

movie house manager in Ohio placed a sign on his theater door: CLOSED—I WANT TO SEE BERLE, TOO!," according to Michael Winship in his book, *Television.*

Puppet shows were a great hit with children. "The Howdy Doody Show" was the most memorable with a live audience of children in the studio. The audience was called the Peanut Gallery. Buffalo Bob Smith, a ventriloquist and puppeteer, was the host. The stars of the show were Clarabelle the Clown and a puppet with red hair and freckles. The puppet's name was Howdy Doody. The show revolved around the ongoing conflict between Howdy Doody and Phineas T. Bluster, another puppet. Like so many TV shows, "The Howdy Doody Show" was an offshoot of radio.

One of the most popular radio shows leaped into television and into a storm of controversy. On radio, "Amos

'n' Andy" was the continuing misadventures of two black men who ran a taxicab company in Chicago. The radio show ran from 1929 to 1954. The voices of the characters were provided by two white men, and that posed a problem for producers of the televised version. They wanted black actors to play the lead roles, but in those days, few were available. The roles for blacks were also very limited when television was young. The parts that were created often portrayed them in a demeaning fashion. The characters were often childlike or gullible, for example. When "Amos 'n' Andy" finally aired on TV, some critics complained that the black characters were negative stereotypes. The show ran only two years before it was forced off the air. Reruns came out many years later, but the show was canceled again amid more protests and controversy.

The Golden Age of Television

"The Golden Age was golden largely in the sense of opportunity. There was an awful lot of drama. Television was still new and exciting. Everybody watched. You could walk down the street the next day and hear people talking about it. You had a sense of an audience, and you had a sense that what you did was needed."

Gore Vidal,
TV writer and novelist

Jack Benny (left) and Bob Hope both had successful radio comedy shows before the arrival of television. As radio listeners became TV watchers, Hope and Benny lost much of their audiences. The two comedians adapted to this situation by moving their shows to television where they again enjoyed success.

Television experienced a period of successful programming throughout the 1950s. This was a time of great creativity and experimentation by makers of TV shows. Writers, production staffs, and performers tested the boundaries, for no one knew just what TV could or could not do in those days. The golden age of TV was not planned, nor was it predictable. It emerged spontaneously from a sense of adventure and fascination with television. Technicians and producers made up the rules as they went along because few had much experience to guide them. Television's potential slowly unfolded before an eager audience. For viewers, television was a magical window to the world, a world of outrageous humor, song and dance, and intense human drama.

While television blossomed, radio lost much of its audience. The ratings of popular radio shows dropped dramatically. Bob Hope, a comedian who had a hit radio show, lost nearly half of his listening audience. The lost listeners had found television. Hope and other performers, like Jack Benny and Ozzie and Harriet Nelson, realized that TV was the entertainment field of the future. Just as audiences shifted their loyalties to the new medium, many radio performers transferred their talents to television.

Television not only had an impact on the radio industry but also affected motion pictures. Americans stayed home to watch TV and went to the movies less often. In 1946, about eighty-two million

Ozzie and Harriet Nelson and their sons, David and Rick, became America's favorite TV family in the 1950s and 1960s in "The Adventures of Ozzie and Harriet." Like other early television shows, the Nelsons' successful TV program had its beginnings in radio.

people went to the movies each week. By 1955, that number had dropped to forty-six million. This caused movie theaters in every state to go out of business. They could not compete with television's line-up of entertainment.

By the end of 1951, the major networks—NBC, ABC, and CBS—had made their first profits in television. For years, they had lost money. By the end of that same year, 30 percent of American families owned eighteen million sets. Television was becoming an important part of the American life-style.

Dramas

During the golden age, dramas filled the nightly TV schedule. Quality dramatic productions were the main reason the 1950s were labeled as the golden age. Audiences were moved and hypnotized by the unending tragedies befalling ordinary people on the screen. They sympathized with the problems that the characters faced and perhaps

felt grateful for not sharing them in real life. But even before television, dramatic shows on radio drew many listeners. They easily made the transition to TV's version of drama.

Many TV dramas were inspired by stage productions. The first TV set owners were generally wealthy citizens who lived in large cities and attended live plays often. Networks wanted to keep this audience happy and decided to develop TV shows with a playlike format. To do this, the networks hired writers, lighting technicians, performers, and others with experience in stage productions. Some of the early TV dramas were actual New York plays that were recreated for television. Though TV borrowed heavily from the stage at first, original dramas made just for television were also produced. "Playhouse 90" was among the leading dramatic shows of the golden age. A ninety-minute, live production aired each week with a whole new set of performers and new hardships.

Because the audience liked them, dramas were a money-maker for commercial sponsors. Large companies wanted to sponsor dramatic shows because they drew audiences who would see their products advertised. The Kraft Company became the first one to sponsor a weekly, sixty-minute drama. It was called "Kraft Television Theatre." From 1949 to 1960, Kraft backed about 650 episodes of the show, many of which were based on great books and plays. One of the most ambitious Kraft productions was the true story about the sinking of the cruise ship *Titanic*. For this show, 107 actors were assembled, thirty-one sets were built, and seven cameras were used for various shots. To add to the challenge, this massive effort was produced on a relatively small stage.

Three scenes from the "Kraft Television Theatre" drama series are reproduced here. Weekly TV dramas used different stories, sets, and casts each week, which amounted to a very expensive production. The Kraft Food Company was the first to sponsor a weekly television drama.

Hallmark Greeting Cards also turned its attention to the world of television. Hallmark sponsored lavish specials, plays, and operas. All of Hallmark's productions were known for their excellence. "Peter Pan," "Pinocchio," and "Arsenic and Old Lace" were just a few of the shows sponsored by Hallmark.

But few dramas could compete with the quality of the Hallmark series. Weekly dramas were expensive. Writers had to be paid for each script, sets designed for each show, and wardrobes fitted for each performer. The cost of weekly drama series may have ultimately led to their downfall. Others believe that the age of drama lost its audience when less

wealthy people started buying sets. They may not have appreciated the quality productions fashioned in the style of big-city plays.

Dramas had other problems, too. No one could be certain about the number of viewers an original story might draw because, from week to week, each story was unique. There was no continuity in plot or in the cast of performers either.

The Birth of the Situation Comedy

The first successful TV situation comedy, or *sitcom,* aired in 1951. Lucille Ball and her husband, Desi Arnaz, starred in the enormously popular "I Love Lucy" show. Lucille Ball first appeared in numerous roles in motion pictures in the 1940s, but fame as a movie star eluded her. In 1948, she changed career paths and starred in a successful radio show called "My Favorite Husband." She played the part of a scatterbrained housewife, a trademark that later made her world-famous on television. Desi Arnaz was a talented Cuban bandleader and singer. When Lucille and Desi teamed up with CBS in 1951, a timeless comedy of unparalleled appeal emerged.

CBS was in the market for a solid sitcom and encouraged Ball and Arnaz to submit their ideas. With the help of writer Jess Oppenheimer, they developed the format for "I Love Lucy." The proposed story focused on the wacky schemes of Lucy Ricardo, a redheaded housewife with a wild imagination. Her husband, Ricky, was a Cuban bandleader who owned a nightclub in New York. Of course, the roles were created specifically for Ball and Arnaz. Each week, the same characters were to appear in

A scene from perhaps the most successful situation comedy series in TV history, "I Love Lucy," shows Lucy performing with two of the show's other stars, William Frawley and Desi Arnaz.

essentially the same family situation. This standardized format was called a situation comedy.

Within four months after the first episode aired, "I Love Lucy" became TV's top-rated comedy. Americans everywhere fell in love with Lucy. In a time of relative economic and political stability, audiences saw much of themselves in the main characters. Lucy and Ricky lived comfortably yet modestly. They watched their budget and for many years did not own a car. Like most women of the day, Lucy stayed home to tend the household and her child. And perhaps also like many, she dreamed of having her own career—in show business. "The believability of all our unbelievable situations is what made it funny," wrote Ball. "People could identify with my zaniness, my

wanting to do everything, my scheming and plotting, and the way I cajoled my husband. People identified with the Ricardos because we had the same problems they had."

The "I Love Lucy" brand of comedy was so well-received that the show was rated No. 1 for four out of the six years that it ran. Though all the lovable stars of "I Love Lucy" have died, they live on in endless reruns. Reruns have appeared on television for over thirty years all over the world.

The Unpredictability of Live Performances

During the golden age, many programs aired live. In a live TV show, the camera records and transmits the show as it happens. Whatever people did during the performance, good or bad, was transmitted over the airwaves. Live performances had their magic moments, but the potential for mistakes overshadowed them. "Stage fright, flubbed lines, accidents on the set—all manner of catastrophe could befall those early days on television—and did," wrote Michael Winship.

Sometimes the camera operator missed a cue or focused on the wrong performer. On one show, the camera mistakenly panned, or moved its focus, to a "dead" body that stood up and walked offstage. In another show, a traveling phone booth stole the scene. In this production, the camera and sound equipment were positioned to capture actor Lee Marvin speaking on the phone. But the live performance did not go according to the rehearsals. Marvin bounded into the booth with such energy that it started to roll.

Because the studio floor was uneven, the booth gathered speed and kept moving. As the booth rolled, so did the camera. It followed every move as the booth traveled across the studio, past an actress who was changing clothes for the next scene, and finally into a wall at the opposite end of the studio. Marvin delivered his lines into the phone during his short trip, but the sound equipment picked up nothing.

Unusual weather conditions appeared on one live TV show. The story called for two actors to sit at a table inside a house while snow fell outside of a window behind them. A stage hand was supposed to sprinkle bits of paper outside the window to make the audience think it was snowing. Somehow, the stage hand became confused, and snow began to fall inside on the actors. The actors continued reciting their lines even though they were soon covered with bits of paper.

Live TV shows required performers to rehearse their lines and actions. Still, that was no guarantee things would go as planned when the show aired live. Actor Lon Chaney often played roles of wild, slightly crazy misfits. On one TV show, he was supposed to lose his temper, pick up a chair, and break it over the head of an actress. The chair was a prop made of very light wood called balsa. Toothpicks held the chair together so that it would break apart easily upon impact. During the rehearsals, Chaney was not supposed to actually break the prop. The big crash was to be saved for the live performance. In rehearsal, he simply walked through the action with the actress and said, "Then she sits, and then I hit her over the head, then I put it [chair] back down." What happened during the live perfor-

mance was described by lighting expert Imero Fiorentino, who was there:

"So now we're on the air. He [Chaney] goes to pick up the chair, but for some reason, he thought he was still in rehearsal. He looks right into the camera and says, 'Here's where I hit her over the head with the chair.' And puts the chair down! The whole control room died. Died!"

Film and Videotape Replace Live Performances

In the 1950s, film eventually replaced live shows. The first TV show to use film was "I Love Lucy." Using film for TV offered many advantages. Mistakes, such as falling scenery or flubbed lines, could be cut out. Film made life easier on the performers because they did not have to rush to change clothes for the next scene or worry about making mistakes. Film also allowed scenes to be shot out of sequence. For example, if an expensive set was built, all the scenes that were to occur there could be shot at one time, even if they were not shot in the exact order according to the plot. After the remaining scenes were shot, they were pieced together in the proper order. None of this was possible with live broadcasts.

Techniques for using film for TV were virtually the same as for making motion pictures. TV film comes on a long roll of plastic that is sensitive to light. For TV production, the film roll is mounted on top of the camera and fed past the lens. As the film rolls, it picks up twenty-four separate frames, or pictures, of the image the camera is focused on every second. This is far too fast for the human eye to notice where one frame stops and another begins. After the show has been recorded on film, it must be processed at a laboratory before it can be aired.

A show produced on film can be shown again and again. These are called reruns, and they are usually aired in the summer months during prime time. Segments of a filmed series can also be sold to local broadcast stations, the cable channels, and to overseas TV markets. Film is still used today for made-for-TV movies, action series, and most documentaries. Local news shows may be a combination of live reporting, filmed inserts, and videotaped sequences.

Like film, videotape records both the picture and sound but has one major advantage. The program can be replayed almost instantly without a trip to the lab for processing. Videotaping requires the use of a TV camera and a videotape recorder (VTR). After the camera picks up the image and changes it into electronic signals, the signals travel through a cable to the VTR. Inside, the signals are changed into a magnetic pattern on videotape. Videotape runs slightly faster than film and captures thirty frames of an image per second, as opposed to twenty-four for film. When the tape is played back, the pattern becomes a series of electronic signals that is transmitted to home antennae on carrier waves.

Television Cowboys

Stories about the old West filled the TV airwaves in the 1950s. By 1959, over thirty-two westerns blazed a trail across television screens of America. But westerns were not new to the public. Audiences

The cast of "Gunsmoke" brought to life well-defined characters with whom viewers could identify emotionally. "Gunsmoke" started as a radio program and then enjoyed a twenty-year run as a hit TV series. The character of Quint (second from left) was played by a young, aspiring actor later to become a movie idol, Burt Reynolds.

had already developed a strong appetite for them thanks to radio, motion pictures, and novels.

The nation's most enduring western got its start in radio before turning into a hit TV series. The show was named "Gunsmoke." Good acting, well-defined characters, and thought-provoking scripts made "Gunsmoke" an enduring weekly event. Written for adults, the show ran from 1955 to 1975 on CBS. James Arness played U.S. Marshal Matt Dillon of Dodge City. Other cast members included Dennis Weaver as the deputy, Amanda Blake as the saloon owner, Miss Kitty, and Milburn Stone as Doc Adams. Marshal Dillon typically tried to persuade the villains not to make trouble. When reason failed, only then did he resort to his gun and fists. Dillon did not always win. On one show he wrestled with the realization that he had hanged a man for a crime that someone else had committed. In another episode, he had to amputate a friend's leg to save the man's life.

American audiences liked westerns and the simple, rugged life-style they portrayed. For a few hours each evening, audiences of the fifties could forget about the threat of communism, nuclear war, and how they were going to pay for their children's college educations. Westerns took these people back to a time when survival was the major goal, and people had fewer responsibilities. Arness explained why audiences identified so closely with westerns: "People like westerns because they represent a time of freedom. A cowboy wasn't tied to one place or to one woman. . . . They're [viewers are] overcivilized. That is why they tune in on western shows, to escape conformity. And they certainly don't want to see a U.S. Marshal come home and help his wife with the dishes!"

By the close of the fifties, there were just too many cowboys riding off into the sunset and fending off the bandits. Audiences had had enough of Hollywood westerns. During each year of that decade, new western heroes were introduced, and it became difficult to separate one from the other. Most of the heroes were tall, rugged, and fast with a gun when forced to use it. Not only did the characters seem too similar but the stories were no longer fresh and original. Sponsors also worried. "With so many western shows on the air, it's difficult to remember which cowboy is plugging what product," wrote *TV Guide.*

"Just the Facts, Ma'am."

Sgt. Joe Friday and his partner, Detective Frank Smith, were the low-key heroes of "Dragnet," a serious cop show. Multitalented Jack Webb played Sergeant Friday of the Los Angeles Police Department. Webb also created and produced the TV show, which began in 1951 and ran for seven seasons. "Dragnet" was the first hit police series on TV. But even before that, Webb had produced the popular radio version of "Dragnet." The success of the radio show encouraged him to try producing it for television. Webb formed a TV production company called Mark VII and convinced NBC to buy his series. The episodes were filmed at the Walt Disney Studios, and each one was shot in about three days. The cost of each show was about thirty thousand dollars.

"Dragnet" was known for its realism. As the producer, Webb wanted viewers to see what police work was really like—tedious, frustrating, and not very glamorous. Legwork and perseverance were the keys to solving most crimes. Webb was also deeply concerned about television violence on the show and allowed no more than one bullet to be fired every four episodes. A man of strong principles, Webb did not want anything on "Dragnet" that he would be ashamed for his own children to see. He also listened to audience reactions to the show. If as few as ten viewers complained about something, he changed it.

Star and producer Jack Webb checks a camera angle during filming on the set of the hit crime series "Dragnet." His TV partner Frank Smith, played by veteran actor Harry Morgan, stands to the left.

The stories on "Dragnet" were based on actual cases taken from the files of the Los Angeles Police Department. Each show opened with the following dramatic announcement: "The story you are about to hear is true. Only the names have been changed to protect the innocent."

Sergeant Friday's no-nonsense approach prompted him to remind witnesses that all he wanted was "just the facts." After the crooks were tracked down, the show closed with a lineup of the guilty parties and the sentence each received. This dramatic conclusion reinforced the idea that "crime does not pay," a recurring theme in the series.

After two hit seasons on the air, shows with similar characters and themes surfaced. "Dragnet" started a trend of law-and-order dramas that continues today. "Pentagon Confidential" was the first show inspired by "Dragnet." This show used files from government agencies instead of those taken from the police. "The Lineup" and many others soon followed. Few of the imitations, however, could match the style and quality of Webb's production. "Dragnet" went off the air in 1959 but was successfully revived for three more seasons in the late 1960s.

Lessons from the 1950s

During television's first decade of growth, the medium discovered its potential. As producers, writers, and technicians tackled new projects, they learned what would work technically and what the audience liked. Audiences were hypnotized by drama, entertained by westerns, and delighted by comedies. Television was eager to please.

One hit show usually led to several more with similar themes and characters. Perhaps if westerns had not become so abundant, viewers would not have tuned them out for the next three decades. But sitcoms were a different story. Audiences never seemed to tire of laughter, no matter how many of these shows aired every night. Even today, they remain a staple of the modern TV diet. Architects of the golden age learned their lessons well. They built a foundation that influenced the TV industry for decades to come. Few TV executives, however, could predict the magic formula that would result in a successful show. Well-written scripts, good acting, and a favorable time slot were necessary, but there were still no guarantees.

The Elements of TV Production

Creativity, technical ability, and teamwork are key elements in all TV productions. Countless hours of behind-the-scenes planning occur before the actors and actresses utter their first lines in front of the camera. Scripts must be written, camera angles set, and scenery and costumes designed. John Rixey Moore, a star on "General Hospital," described the production teamwork on his show: "It's an ensemble effort with forty to fifty people present behind the scene. If someone flubs his or her lines, everyone has to stop."

The production staff and technical crew work as one unit. Every move is coordinated with precision timing to effectively tell the story in pictures and sound. The creative part of the TV team is commonly known as the production staff. The production staff consists of the producer, director, writers, and assorted assistants. The technical director, audio engineer, lighting director, camera operators, and stage manager make up the technical team.

Though members of the technical team must have specific knowledge to operate complex equipment, they also need a sense of creativity to complement their technical abilities. The lighting staff creates a dramatic mood by adjusting both focus and intensity of the lights and by placing them in certain positions. Sound technicians add just the right music to a scene to create either excitement or sadness.

To be effective, the creative team must also possess some technical knowledge. An effective director needs to know the functions and limitations of the cameras, microphones, and lights. Set designers must be aware of how the camera picks up details and how the performers will look in front of the set. On some talk shows, specially designed sets make the host appear larger than life. This is done by seating the host behind a desk that is two-thirds the size of a normal one. As the host towers over the desk, he or she fills up the camera shot.

Some of the Key Players

The producer is perhaps the most important person behind the scenes of a TV show. He or she is responsible for the entire production from start to finish. If the show is a new sitcom, for example, the producer develops its concept, or theme, and the characters. The producer works closely with the script writers to create entertaining stories. For a series, the characters must be colorful, have unique personalities, and face a new challenge on each show. Managing the budget and administrative matters are also part of the producer's job. Because of so many responsibilities, he or she usually has several assistants and delegates much of the work to them.

The director works for the producer. He or she is in charge of the overall look, feel, and sound of a show. The goal of the director is to fulfill the producer's vision of what the show's style and content should be. Offstage in a control booth, the director signals the stage manager when to roll cameras and when to break for commercials. The stage manager gives this cue, an instruction to begin action, to performers and camera people. After all of the scenes have been filmed to the director's satisfaction, he or she supervises the final editing process.

Offstage, the technical director (TD) operates equipment in the control booth. The most important piece of equipment there is called the switcher, a large console of buttons, controls, and monitor screens. Pictures from different cameras appear on the monitor screens. The director and TD select which picture will go out over the air or will be recorded for later viewing. Sometimes the best directors double as their own TDs.

Sound for TV productions is controlled by the audio engineer. In a special sound-control console, the engineer mixes the voices of the performers with prerecorded sounds or music. The sounds must be balanced so that the voices of the performers are not drowned out. Pre-recorded sounds are sometimes added and coordinated with the action. These are created by audio experts who are able to duplicate, for example, the sounds of shattering glass, gunshots, and footsteps on carpets. With the help of the audio crew, the audio engineer directs the positioning of the sound equipment, such as the boom microphone. This microphone is suspended on a long mechanical arm that can be moved around over the heads of the actors to pick up their voices. The boom microphone is not

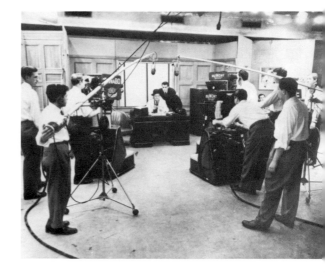

(Top) Producing a TV program usually calls for more people behind *the scenes than it does actors* in *the scene, as this picture shows. Members of both the production staff and the technical team have important roles in creating a TV program. (Bottom) Members of the "Peanut Gallery," the studio audience of the "Howdy Doody Show," get a camera's-eye view of the popular 1950s children's program.*

supposed to appear on camera, but occasionally it falls too low into a scene. If the director does not notice it, viewers will see the microphone on their screens.

Lighting can convey a variety of moods on the TV screen and is an important element in the visual impact of a show. Effective lighting can make performers look better, too. The lighting director is responsible for this essential production function. He or she also ensures that the cameras have the correct level of illumination to create the particular mood that the director wants. Subdued lighting may create a feeling of suspense or romance. Bright lights may be needed for a party or restaurant scene.

In the production studio, lights move above the sound stage on a huge horizontal structure suspended from the ceiling. This structure is called a scaffold. The scaffold light-suspension system is constructed in a crisscross grid that looks somewhat like graph paper. Each light position on each part of the grid is labeled and coded so that it can be easily referenced for adjustments. The electrical technician who works with the lights is called a gaffer.

Road Maps for the Production

Just about every type of television show has a script or an outline of events called a run-down sheet. Scripts and run-down sheets are important because they provide a structure for organizing the show. They are the road maps that the crew and performers must follow. Without a structured approach, the production effort would be disjointed and would also require more time and money to produce.

Though talk shows and quiz shows cannot be scripted word for word, the flow of events has to be organized and timed. This information is organized on the run-down sheet. The sheet simply lists the person who will do the introduction, news, or interviews. Sometimes the introductions that the talk show host needs to read will appear on a TelePrompTer, a device that unrolls a magnified script in front of the speaker. He or she can look into the camera and

NBC vice-president Jules Herbuveaux and TV star Kay Westfall pose in 1956 at the first TV studio to be outfitted for all-color telecasts. The studio belonged to the NBC Chicago affiliate WNBQ. On April 15, 1956, the NBC network became the first all-color TV network in the nation. Notice the huge banks of lights on the ceiling. Color television requires three times the light needed for black-and-white television.

at the same time read the text as it scrolls by. For talk shows and quiz games, each segment is carefully timed with commercials placed throughout.

The Role of the Script

For drama and comedy series, the script is the TV story told through dialogue and action. A strong script is critical to the success of a show. Even if the show is already a hit, a weak script can influence viewers to turn it off, causing ratings to drop. Sponsors pay more money to advertise on shows with high ratings, so an unappealing script can translate into less money for the show.

Performers bring the script to life. Facial expressions, voice tone, and body language are the actor's tools. "Learning my lines is a great priority. I usually have a week in advance to learn them. The problem is making them believable," said John Rixey Moore. Well-trained performers can "pop in and out of their emotions with ease," he said.

Like books or magazine articles, scripts have a unique format to make them easier for performers to read. This format consists of wide margins and a lot of space between each character's lines. The extra space on the page allows technicians, actors, and the creative staff to make notes on when or how they are to perform their jobs.

The story the script tells often unfolds in formula fashion. For a weekly sitcom or prime time drama, conflict must be introduced within the first few minutes of the show. By the first commercial break, the audience should be sufficiently enticed to watch the rest of the show. They should want to see how the characters get out of their troublesome situation.

A script is composed of individual scenes that propel the story forward, reveal aspects of the characters' motivations, and develop relationships. "One basic overriding rule for successful television is it's got to have appealing relationships or else it doesn't work," said Tim Brooks, author of *The Complete Directory to Prime Time Network TV Shows.* By the end of the show, problems must be addressed, relationships stabilized, and the mysteries solved.

The Cost of TV Shows

The cost of producing a weekly TV series is staggering. An hour-long prime time show can run $1 million per week. A half-hour prime time show costs about half as much or sometimes more. The hour-long news shows, such as "20/20" and "48 Hours," are much cheaper to produce. They do not cost as much primarily because they use a single, permanent set and because the producers do not have to pay people for interviews or for their stories.

Like any other industry, salaries are a big part of the TV budget. For a weekly show, stars can receive many thousands of dollars for each episode. Wages are frequently set by unions, which represent the various professions in the production team. These include the actors, directors, and musicians. Top performers and news personalities, like Dan Rather, can receive several million dollars per year. Even people who appear as just faces in a crowd must be paid. They are called extras. Other expenses include props, transporting equipment to a location outside the studio,

The cavernous studio at CBS's "Television City" in Hollywood seems to dwarf the set being televised. Huge monitors hanging from the ceiling show how the telecast looks to the TV audience. Paying actors and technicians, and buying or renting equipment, accounts for a large part of the cost of producing a television show.

and processing film. The list of expenses actually runs several pages. Keeping expenses within budget is a challenging task, even for the most experienced producers.

Putting It All Together

A single television program is the result of the combined efforts of many talented people. Members of the television team must be masters of their craft. In addition, they must be able to coordinate their efforts and produce an end product that will please the most important members of the television team—the audience.

Much of the work of putting together a TV program takes place in a TV control room like the one pictured here. Today's control rooms use state-of-the-art technology, particularly computers, to get the job done. High-tech equipment has enabled TV producers to set up mobile control rooms which allow for live, on-location telecasts.

Television and Advertising

Television has become one of humanity's most popular inventions. Its ability to entertain and to inform has helped place it in over 160,000,000 homes in the United States. But television may do more than entertain and inform. According to many experts, television can also persuade. These experts believe that through commercials, television can influence consumers to buy certain products.

Prior to 1941, no commercials aired on TV. The FCC considered television to be experimental and would not allow TV to sell airtime for product advertisements. As a new industry, television was operating at a loss, with no profits in sight. Selling commercial airtime was the way that television executives planned to make back the money spent on bringing this invention to the nation.

NBC and David Sarnoff urged the FCC to grant the network commercial status. On July 1, 1941, the FCC finally agreed. Both NBC and CBS, the two major networks, were granted commercial status. But certain conditions had to be met.

First, television stations had to broadcast a minimum of fifteen hours per week. Experimental TV programming had been unpredictable, with large gaps between shows. Second, stations had to transmit 525 lines of electrical impulses to TV sets, ensuring a good-quality picture for consumers. In return, the networks were authorized to sell airtime for commercials, a venture that would ultimately yield vast fortunes.

Compared to the excitement of modern commercials, the first commercial on TV was very simple yet effective. The face of a watch made by the Bulova company ticked on the screen for sixty seconds. Bulova paid only nine dollars to air the commercial, a tremendous bargain compared to current advertising rates. The commercial aired on NBC's station in New York City, where about four thousand TV sets were in use.

By the mid-1940s, powerful advertising agencies in New York City were taking TV commercials seriously. Though most of their ad campaigns had been prepared for radio and print media, they imagined an exciting and profitable role for themselves in television advertising. The head of one large ad agency, J. Walter Thompson, predicted that television was to become "the biggest ad medium yet."

Thompson helped make his own prediction come true. He teamed up with the Kraft Company to sponsor "The Kraft Television Theatre" and, of course, to sell Kraft products. From Thompson's point of view, the show was a testing ground to determine TV's ability to sell. The product that they advertised was McLaren's Imperial Cheese, a new item with an expensive price tag. At one dollar per pound, the cheese did not sell well. In the commer-

An advertisement makes a visual connection between wearing a Bulova watch and being an American. The ad appeals to viewers' feelings of patriotism and their desire to be seen as good citizens. These feelings and desires can be fulfilled, the ad hints, by owning a Bulova watch.

cials, an attractive female model sampled the cheese while an announcer commented on how good it tasted. After running the ads for just a few weeks, consumers were convinced that McLaren's Imperial Cheese was just what they needed. The cheese disappeared from grocery shelves. Successful marketing of the cheese proved that TV was a powerful persuader, and the new advertising medium passed its first test with top scores.

The Great Persuader

TV advertising is the ultimate salesperson. Just one commercial can introduce millions of viewers to a new product or service or remind them about a familiar one. Ads try to help consumers see how the product fits into their life, solves problems, or benefits them. Often, a commercial creates a need that viewers may not have felt until after they finished watching it. By "hooking" the consumer mentally, ads can reel in consumer dollars at the cash register.

Television advertisers use many approaches to get the public's attention. With less than sixty seconds of airtime, products in the ads have to convince the consumer to "buy me." Many techniques are employed to get the message across to viewers. These include the use of celebrities or athletes to endorse the product, loud music, young women in bathing suits, special effects, or humor.

Looking for new ways to make a sale, several large companies turned to famous celebrities and sports figures. In 1984, Pepsi-Cola paid superstar Michael Jackson and his band several million dollars to make its commercials. Young people in particular like Jackson's music and respect him. As a result, Pepsi sales jumped. Nike, a manufacturer of sportswear, featured a popular commercial with football player Bo Jackson and musician Bo Diddley.

Special effects in a Hertz Rent-a-Car ad helped the company capture 70 percent of the rental-car market in the 1960s. The ad showed a man in a business suit flying through the air and

TV star and sex-symbol Don Johnson of "Miami Vice" endorses Coca-Cola in this advertisement. Advertising agencies know that Americans love to imitate TV heroes and will buy a product just because a celebrity uses it—or at least appears to use it. Whether or not Johnson drinks Coke does not matter. The important thing is that people connect it with him.

landing in a Hertz car. The announcer said, "Let Hertz put you in the driver's seat," and many people did.

In the mid 1980s, humor in an ad for Wendy's restaurants hooked Americans all over the country. The commercial made audiences chuckle while making the competition look cheap. The idea was simple. An elderly but assertive woman ordered a hamburger from another restaurant. When the order arrived, she demanded to know, "Where's the beef?" The ad implied that there was no shortage of beef at Wendy's. The audience got the message, and Wendy's revenues swiftly jumped by 31 percent.

Not all advertisements on TV hit the mark. Pepsi-Cola learned that famous people in their ads hold no guarantees of public acceptance. Like Michael Jackson, singing star Madonna was paid several million dollars to appear in a Pepsi commercial. But the commercial included religious references and symbolism that offended some people. Not wanting to upset or lose customers, the company took the commercial off the air after running it only one time.

Superstar athlete Bo Jackson jams with blues legend Bo Diddley in a commercial for Nike athletic footwear. The ad appeals to a wide audience of both sports and music lovers and has given the phrase "Bo knows" a place in popular speech.

Behind the scenes, sponsors use their powers to persuade in another way. They have a voice in the produc-

They don't send you out of town to fail

But success doesn't come easy.

Ask our friends on the right. They'll tell you this: when you're out of town you have to sell, make decisions, solve problems.

To do this you have to keep your mind on your job. Not on ours.

So we make sure you get a frisky, flawless Chevrolet or other fine car that you never have to think about. You should be distracted by profound considerations of gas tanks, empty or full—wipers, working or wacky? Ridiculous!

Busy, successful men are sent out of town to—succeed. That's why we try to make every Hertz office a way station of success.

Your success.

Russ Jacobson—London Fog

Jim Roth—Woolco Stores

Bill Heubach—Union Bag-Camp Paper

Let Hertz put you in the driver's seat

(ISN'T THAT WHERE YOU BELONG?)

Special effects in an advertisement for The Hertz Corporation helped the company capture 70 percent of the rental car market in the 1960s. The aim of a TV advertisement is to persuade the viewer that his or her needs, dreams, and ambitions can be fulfilled by the sponsor's product. In the Hertz ad, for example, the viewer's ambition for business success is targeted.

tion of TV shows, what shows are backed, and what the script looks like. Sponsors want to "play it safe" and appeal to as many viewers as possible. After all, if viewers turn off the show, they will not see the commercials. Sometimes a sponsor will cancel a commercial on an ongoing series if the script is poor or one of the stars gets into trouble, such as getting arrested for drugs or posing nude for a magazine.

What Are They Really Selling?

Products sold on television have images and personality. They are the stars of the commercials, and viewers must somehow be convinced that their lives are incomplete without the product. On TV, a bar of soap does not just clean but invigorates, gives the user self-confidence, and attracts admiring

Ads for Wendy's restaurants, featuring actress Clara Peller, pictured here, made a mark on popular culture with her memorable portrayal of a patron who bellowed "Where's the beef?" when she got her hamburger at one of Wendy's competitors.

has become a mass-marketing forum. Advertisers want consumers to feel that their merchandise, no matter what it is, will make consumers happier, more glamorous, more successful, younger, or thinner. The creation of these illusions has developed into an art form in the advertising world. How well advertisers conjure up the fantasy often translates into millions of dollars worth of sales.

Creators of commercials may try to make viewers feel inadequate if they fail to use their products. The fear of bad breath, dandruff, and wrinkles drives Americans to spend over $25 billion dollars a year just on personal care items. This is more than the rest of the world combined spends on such items.

Mothers and wives are often the targets of commercials. The underlying message is that well-informed women who care about their families will select this particular brand of peanut butter or detergent. Advertisers try to convince wives and mothers that they are responsible for fighting "ring around the collar" and winning their family's love by using the right cake mix. One

glances and whistles from the opposite sex. One soap ad implies that those who do not use the product can be detected in a crowd.

Commercials are carefully crafted to coax consumers to buy a variety of things. This is called salesmanship. From used cars to denture cream, TV

Members of an advertising agency brainstorm during a planning session in the early years of television advertising. Businesses were quick to recognize the possibilities for reaching millions of people through the medium of television. Television advertising is now a multibillion-dollar-a-year business.

laxative, Phillips's Milk of Magnesia, abbreviates its name as M.O.M., another attempt to persuade mothers that buying this product is the caring thing to do.

Product Promotion and Misleading Advertising

A certain amount of product exaggeration is generally tolerated under the law and is known as "puffing." Sometimes, however, advertisers make claims that are obvious distortions or use demonstrations that are deceptive. Misleading commercials generally make the product look better in an attempt to persuade consumers to buy it.

During the 1960s, the Colgate-Palmolive Company found itself in serious trouble for saying one thing and doing another in a commercial. The company advertised Rapid-Shave Cream and told viewers that it was so effective that the cream could shave sandpaper, the toughest test of all. Instead of using sandpaper though, advertisers used Plexiglas, a hard, clear plastic, and covered it with sand. Viewers thought they were seeing an impressive demonstration: the razor easily glided across the surface.

The Federal Trade Commission (FTC) was not pleased and accused Colgate-Palmolive of deceptive advertising. The FTC is the government agency responsible for policing unfair or deceptive actions in the business world. The Supreme Court finally ruled on the case and forced advertisers to be more honest in their product demonstrations. The court said that no TV demonstration can use a material that pretends to be something else unless the audience is informed. The court also required that visual demonstrations of products must be truthful.

The FTC has ordered several companies to stop practices that made their products look better on TV. In a commercial for Campbell's vegetable soup, marbles were placed in the bottom of the bowl to force the vegetables to surface. This made the soup appear heartier than it was. To help coffee look richer and darker on camera, one company substituted oil in place of the coffee. Makers of a mouthwash bragged that it killed germs that caused colds, but the government disagreed and forced the company to stop making that claim.

Drug Commercials

The techniques used in ads can influence people to buy the product being advertised, but these ads can also influence people in other ways.

Commercials can persuade viewers of all ages, for example, to say "yes" to over-the-counter drugs. "Drug companies spend hundreds of millions of dol-

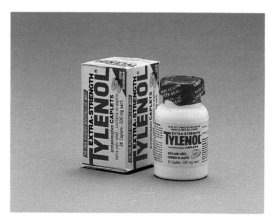

Advertising influences people's outlook on life as well as persuading them to buy a product. Some people believe, for example, that ads for over-the-counter drugs, like the one shown here, encourage an attitude of turning to drugs to escape life's pain.

lars each year to persuade Americans that the solution to every ailment from sniffles to sleeplessness can be found in a pill," writes author Donna Cross in *Media Speak.* Even though no known cure exists for the common cold, hundreds of cold remedies line store shelves and home medicine cabinets. The actors and actresses in cold commercials always appear happier and better able to cope after taking over-the-counter drugs.

These ads can send a false message to impressionable viewers—that pills and medication can make life better. In *Media Speak,* a doctor who worked with drug addicts said, "TV advertisers are teaching our kids to use drugs. . . . I know of no drug except heroin or morphine which will produce the dramatic relief from worldly cares TV vividly pictures."

Commercials for Children

Commercials directed toward youthful consumers abound on Saturday morning television. Advertisers want to sell children dolls, building blocks, cereal, party dresses, and countless other products. On some shows, clusters of up to five short commercials appear at a frenzied pace. Amazingly, a child may see between twenty and forty thousand commercials in just one year.

Advertising agencies are hired by companies with products or services to sell. Companies pay ad agencies roughly $500 million a year to develop commercials that will entice young viewers to spend money. Apparently, the ads are working. According to an article in the *Atlanta Constitution* newspaper, "children between the ages of 7–11 spend $5

billion a year and influence another $50 billion in family spending." If they cannot afford something themselves, youngsters convince adults or older siblings to make the purchase.

But adults continue to be the main target of TV advertising, and that has hurt children's programming. Many advertisers do not want to sponsor children's shows because children do not have the tremendous buying power of adult consumers. This economic fact has caused a major reduction in children's shows, especially during the week. "Children's programming of *any* kind, let alone of high quality, has become almost invisible on weekdays outside of public television and independent [non-network-affiliated] commercial stations," writes Edward Palmer, author of *Television and America's Children.*

Critics of TV advertising consider some children's shows to be just one long commercial. These product-related programs, such as "My Little Pony" and "Teenage Mutant Ninja Turtles," feature the toy itself as the dynamic star of the show. They are "toy promotions disguised as stories." The problem is that there is no clear distinction between the story and the advertising.

But product-linked programs are not new. In the 1950s, a favorite, high-quality family show was also a low-key sales pitch. According to David Diamond, writing in *TV Guide,* "When the long-running and highly acclaimed television program 'Disneyland' premiered on Wednesday night, Oct. 27, 1954, few viewers—young or old—knew that it was in part a marketing tool to promote the theme park of the same name that was being built in Anaheim, California."

An advertisement for Disneyland promises fun and fantasy. A 1954 TV program called "Disneyland" was actually an advertisement in disguise for the theme park then being built in Anaheim, California. Watchdog groups who monitor the quality of children's TV programming say that many children's programs are just long advertisements masquerading as entertainment.

One organization, Action for Children's Television (ACT), has kept a close watch on ads and programming for young viewers. Since 1968, ACT has fought to eliminate commercial abuses and encourage more choices in children's television. The organization was established by four concerned mothers and has taken legal action on many issues that affect children's TV. ACT, for example, has worked to reduce the number of commercials during children's programs.

Who Pays for Commercials?

Making a commercial is expensive. Companies with a product or service to sell hire advertising agencies to make the arrangements. Ad agencies develop the sales approach, locate performers, and film the commercial. These costs can run into the millions. After the commercial is made, the sponsors must also buy airtime, another big expense. The cost of airtime depends on a number of things—how long the commercial runs, what show it sponsors, and when it is shown. Ads shown on a hit, prime time series are generally the most expensive, costing an average of about $300,000 per minute. Hit shows lure a much bigger audience, more potential customers, and thus more potential profits. In 1990, NBC is expected to earn over $350 million advertising dollars. ABC is expected to make $150 million and CBS, $85 million.

The cost of TV commercials has skyrocketed over the last decade. For the

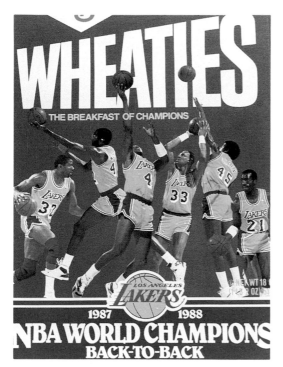

Advertisers try to identify their products with winning images. The cereal pictured here, called "the breakfast of champions," identifies itself with the Los Angeles Lakers basketball team by putting the 1987 NBA champions' picture on the box.

crease. One observer notes that from the advertiser's viewpoint, commercials do not interrupt programs but that programming interrupts the airing of commercials. Industry insiders add that the networks want a hit show, not because it pleases the audience but because they can charge more for selling commercial airtime. Writer Anne Rawley Saldich describes the arrangement this way:

> "It's a simple little system. Broadcasters use our airwaves at no cost to themselves. Consumers (not just people who own television sets) pay for them to have this opportunity. Manufacturers use our airwaves at no cost, also, because they have passed the advertising bill on, and it is from that fortune in advertising revenues that broadcasters make their enormous profits."

1990 Super Bowl, ads sold for a record $1.4 million per minute, or $700,000 for thirty seconds. In contrast, Super Bowl sponsors paid $550,000 per minute in 1980. Because of the increased costs, many companies purchased smaller segments of time during the 1990 game.

In a way, networks sell viewers to advertisers instead of selling shows to viewers. A successful, prime time series draws millions of viewers week after week. That means more people will see the commercials of the companies that sponsor these shows. The product being promoted will therefore get much exposure, and product sales will in-

In the end, the consumers pick up the bill for all product advertising. The cost is tacked on to the retail price of goods and services. But the fact that consumers are persuaded by advertisers to try a product does not guarantee the product's success. Few people will buy a product again if it does not perform well, no matter how attractive the ads. The refusal to repurchase can mean major financial losses for companies. They spend a great deal of money designing, testing, advertising, and distributing products. If the item fails to compete with similar products, a company can incur severe financial losses. Consumers have the ability to knock a product out of the marketplace by simply ignoring it. They too have the power to persuade.

Television and Presidential Politics

Americans rely on television as their major source of political news. They have come to trust television as a news source more than radio, magazines, or newspapers. Without television, some viewers would learn little about political and world events. Research has shown that people with less education and low economic status rely most heavily on television as perhaps their only source of the news. For them, it is an inexpensive way to keep informed of major events.

One reason that viewers trust TV is because they trust the journalists who report the news. Former newscaster Walter Cronkite was highly respected as a reliable source of news by most Americans. He was calm and deliberate in presenting news events. Before he retired from "CBS Evening News" in 1981, he held the title of "Most Trusted Man in America."

Of all the communication devices, television plays the biggest role in covering political developments. Television is important because it gives candidates national exposure within a short time period. During a campaign, it is impossible for candidates to travel to every town and city looking for voter support. But a few well-placed ads can reach millions of people with the candidate's message. Yet despite its tremendous influence, the medium is also limited. Time constraints rarely allow in-depth TV analysis of a candidate's background or the issues.

For years, television news journalist and anchorman Walter Cronkite brought news of important events into the homes of millions of Americans each evening. His style and presence earned him the title of "Most Trusted Man in America."

Presidents and other politicians know that people trust TV newscasters and journalists, and politicians try to maintain good relations with them. Because dealing effectively with the media is important for selling ideas to the public, new presidents hire a press secretary to help them. Usually a former journalist, the press secretary's job is to explain the president's policies, make public announcements, and arrange press conferences with reporters.

Sometimes the press secretary works to change or improve the president's image. But one former press secretary believes that too much emphasis is placed on image-making instead of where the president stands on political issues. "But the fact is when you see a political leader on television today, the whole thing is merely a performance," said George Reedy, former press secretary under Lyndon Johnson. Press conferences especially are highly staged events. Presidents spend hours, sometimes days, rehearsing answers to possible questions that TV and newspaper reporters may ask.

Television and Politics—An Enduring Relationship

Ever since television turned into a household word in the 1950s, politicians discovered that it could help them get elected (or defeated). It could also help them promote their policies once they were in office and allow them to dominate the news. Presidents and aspiring presidents learned they could influence large numbers of people through the electronic wizardry of television. NBC newsman John Chancellor commented about the present relationship between TV and the president. "Presidents of the United States in the age of television are media monarchs [kings] with access to everybody's home television set just about anytime they want it. This monarchial power has been overused."

After World War II, television expanded throughout the country. As more families bought TV sets, politicians realized they could reach out to them through the airwaves. In 1947, President Harry S. Truman was the first president to televise his state of the union address, an annual report of the nation's progress. Later during his term, Truman made short speeches for television audiences from his office. Awkward and untrained in performing

President Lyndon Johnson interrupts his press secretary George Reedy during a briefing that Reedy was conducting in 1964. Politicians, like businesspeople, quickly learned that television can create a favorable or unfavorable public image.

before the camera, Truman simply read from a notebook, looking up at the camera from time to time.

Truman's successor, Dwight (Ike) Eisenhower, used the new medium in a more aggressive fashion—to help him win the highest office in the land. In 1952, Eisenhower and his vice-presidential running mate, Richard Nixon, helped bring TV into the political arena. Eisenhower was uneasy in front of television cameras, but he realized he could reach many voters via the new medium. To help sell himself to the public, he hired an advertising agency to produce short commercial spots on his behalf. Some of the commercials featured bands marching to the catchy slogan of "I like Ike. You like Ike. Everybody likes Ike." Eisenhower's opponent, Democrat Adlai Stevenson, relied on televised speeches, instead of short spots, to get his ideas across to the public. Stevenson lost the election. After his defeat, Stevenson is said to have remarked, "The idea that you can merchandise candidates for high office like breakfast cereal . . . is the ultimate indignity to the democratic process."

As president, Eisenhower wanted to look and feel more confident on television. To accomplish this, he consulted an expert in the television business, Robert Montgomery. As the successful producer of his own show on NBC, Montgomery was experienced in making performers look good in front of the camera. He coached Eisenhower on the finer points of performing and how to act natural in front of a camera. A professional makeup artist and rimless glasses improved Eisenhower's TV appearance. Montgomery also experimented with different camera angles and lighting to show the president's best features.

Dwight D. Eisenhower was the first presidential candidate to exploit TV as a campaign tool in 1952. He hired an advertising agency to produce short commercial spots for him. After Ike won the election, his opponent, Adlai Stevenson, criticized him for marketing himself like breakfast cereal.

The TV Camera—Political Friend and Formidable Foe

As the 1952 vice-presidential candidate, Richard Nixon proved no amateur when it came to making the power of TV work for him. Nixon is considered "among the first politicians to comprehend and benefit from the skillful use of television." He proved it when his nomination to run with Eisenhower came under fire. Just after he was nominated, outcries of improper behavior erupted. To plead his case, he turned to TV audiences in a nationwide appeal.

Presidential candidates John F. Kennedy and Richard M. Nixon confront each other on the issues during the first televised presidential debate in 1960. In the broadcast, viewers thought Kennedy won the debate. People who listened to the debate on radio, however, gave the edge to Nixon.

In a humble speech, Nixon admitted accepting financial gifts. He explained that the money was used so that he could better serve his supporters. The next part of the speech had nothing to do with the charges but helped viewers see that he was just an average person like them. He spoke of his boyhood and about how his family had been poor. As an adult, he had a home mortgage, a two-year-old car, and a modest life-style. Nixon ended the speech by admitting he had accepted another gift—a little black-and-white cocker spaniel. His young daughter had named it Checkers. Nixon announced that he planned to keep this gift no matter what.

Richard Nixon plays with his dog Checkers. Nixon's skill in appealing to the public's sympathy via a televised speech in which Checkers appeared thwarted an attempt to reject Nixon as Eisenhower's running mate in 1952.

An outpouring of support resulted from the now-famous "Checkers" speech. The public flooded Republican headquarters with telegrams defending Nixon. They wanted him to stay on as Eisenhower's running mate. Nixon went on to serve two terms as vice-president, thanks to his masterful ability to use television as a campaign tool.

Although TV worked to Nixon's advantage at first, his long-term relationship with the medium was stormy. Television eventually contributed to his political downfall. Clouds appeared in 1960 when he ran against Democrat John Kennedy for the office of president. In the first televised presidential debates in history, the candidates went before the TV cameras on four separate occasions and argued their points.

A series of unplanned events detracted from Nixon's TV debate debut. Though the lighting was carefully adjusted to Nixon's advantage before the show, a crowd of photographers clamored into the studio and upset the lighting arrangement just before air time. The backdrop on the set behind the candidates turned out to be a lighter shade of gray than expected. This caused Nixon's gray suit to fade into the background. Kennedy, on the other hand, looked crisp and sharp in a dark suit and tanned skin. Still recovering from an illness and a hard day of campaigning, Nixon looked awkward and tired on the screen. When Kennedy spoke, he looked into the cameras at the audience. Nixon addressed his remarks to Kennedy on the stage. "And most of all, his [Nixon's] advisers now insist, he lacked the energy to project—for Nixon does best on television when he projects, when he can distract the attention of the viewer from his looks to the theme or the message he wants to give forth," wrote Theodore H. White in *The Making of a President 1960*.

Some observers believe that Nixon's TV appearance cost him the election by providing Kennedy with just enough votes to narrowly defeat him. Smooth, confident, and projecting an

John F. Kennedy and Richard M. Nixon are photographed together during the 1960 presidential campaign. Commentators credited Kennedy's later victory in the election to the robust image that he projected on television. Nixon, on the other hand, appeared tired and colorless.

appropriate presidential aura, Kennedy convinced television audiences that he was the winner of the debates. Radio listeners also tuned in to the debates but came away with two different impressions. Lacking the visual message of TV, radio audiences did not consider Kennedy the winner. They felt that either both candidates were equally strong in the debates or that Nixon had come out ahead.

Despite the political outcome of the debates, television was praised for giving the public the opportunity to see and hear the candidates discuss the issues. Through the eyes of the camera, people all over the country observed an important historic event. This ignited more voter interest in the electoral process. Voters turned out in record numbers for the election. Television undoubtedly played a role in bringing people to the polls.

By November 1968, Nixon was back in the running for the office of president. In this campaign, Nixon was determined once again to make television work for him, not against him. With the help of his staff, carefully staged discussions with citizens were set up at campaign stops across the country. The discussions were planned to make Nixon look professional and informed. Still stinging from his bad experience with Kennedy, Nixon refused to debate Democratic opponent Hubert Humphrey.

Instead, the candidates answered questions that were phoned in to TV networks by the viewing audience. Nixon responded to calls on NBC and Humphrey fielded calls on ABC. This arrangement made it harder to compare candidates side by side because they did not appear together on the same channel. Nixon's strategy worked.

Presidential candidate Hubert H. Humphrey and his wife Muriel hit the campaign trail in 1968. Humphrey could not get his opponent, Richard Nixon, to debate him on TV because Nixon felt a televised debate would favor Humphrey, just as it had favored Kennedy in 1960.

The next day, he narrowly defeated Humphrey and became president. Remembering the 1960 debates, Nixon made television work to his advantage this time.

Television Contributes to the Unseating of a President

Though Nixon won a second term as President in 1972, a scandal of historic proportions was brewing. In the summer months before the election, a burglary took place at the offices of the Democratic National Committee in the Watergate office complex in Washington, D.C.

The burglary turned the eyes of the nation and the media toward the White House. At first, CBS was the only network that reported the event. After seeing the first of two scheduled fifteen-minute reports on the evening news, a presidential aide tried to pressure CBS into killing the follow-up story. Though CBS aired the second segment, they cut the length in half.

By the spring of 1973, alarming accusations prompted government action. A Senate committee was established to investigate the burglary and discovered that high-ranking White House officials were involved. Clearly, an important event was unfolding with many puzzling questions to be answered. Television networks took turns covering the investigations. Testimony revealed possible wrongdoing at high levels of the government.

Americans were intrigued with the Watergate broadcasts, and the networks increased their coverage. When a former Nixon aide revealed that the president had an elaborate system for taping conversations and meetings, investigators wanted the tapes. Nixon refused.

By this time, the public was rapidly losing confidence in the president, though no hard evidence had surfaced against him. His popularity plunged, and people talked of impeachment.

Although television brought all the bad news about the White House to the public, Nixon tried once again to use it to turn the tide in his favor. With live coverage by all the networks, he embarked upon a speaking tour around the country. In front of sympathetic audiences and TV cameras, he announced, "I am not a crook," and "One year of Watergate is enough." But this time, the public refused to forgive him, no matter how emotional his speeches.

In a televised speech to demonstrate that he had nothing to hide, Nixon turned over a huge stack of transcripts from his private tapes to the media and to the Senate investigators. Nixon felt the tapes would free him of any blame in the Watergate scandal.

But the plan backfired. When portions of the transcribed tapes were read

President Richard Nixon delivers his resignation speech on TV on August 9, 1974. Television coverage of the Watergate scandal, and Nixon's involvement in it, forced the president into resigning his office rather than face possible impeachment proceedings. Television proved its value in this case as a tool for informing the public of unethical acts committed by elected officials.

on television by all of the networks, Nixon's image was hurt even more. The tapes revealed a "petty, self-centered man with little concern for justice," according to writers Harry Castleman and Walter Podrazik.

A congressional committee seriously discussed impeaching the president when Nixon released yet another tape. This one convinced many more people that he was involved in hampering the investigation of the Watergate break-in. Rather than face further disgrace, Nixon resigned from the presidency on August 8, 1974.

Broadcast live over all the networks, Nixon's announcement offered no detailed explanations. The next day, in a parting, tearful speech before his staff and TV cameras, he spoke of his family and middle-class background and begged for understanding. "It was the Checkers speech all over again," wrote Castleman and Podrazik. Nixon had skillfully used the power of television to ride to the peak of his political career He was unable, however, to harness the same power to help stop his downward spiral.

Television brought the Watergate drama into homes all over the country. Americans had front row seats as they watched the events unfold. Through the unblinking eye of the camera, viewers were fed visual information upon which opinions were made and refined. "Dramatic pictures can give the brain a jump-start and then accelerate the opinion-formation process," said Philip Seib in *Who's in Charge*. Television helped convince the public that something very wrong was going on in the White House, though all the facts never came out. Once more, while faith in the president was dying, public trust in television as a news source was growing.

President Ronald Reagan announces to the nation during a June 15, 1987 televised speech that he will propose nuclear arms reductions in Western Europe.

The Reagan Years

Former president Ronald Reagan has been called "the made-for-television president" and "the great communicator." His reputation was well-deserved. During his two terms in office from 1980 to 1988, Reagan displayed a unique gift for communicating ideas on television and radio.

Reagan spent most of his early life learning the art of media communication, first as a well-known sports announcer and later as a movie actor. He also hosted and sometimes starred in a popular TV drama show, "General Electric Theatre." Reagan also served eight years as the governor of California. Many believe that his experience and training as an actor helped him a great deal after he went into politics. He knew how to deliver a line, whether it was from a movie script or a political speech.

President Reagan waves to onlookers on his way to boarding a helicopter on the White House lawn. Some experts believe that Reagan used television as a political tool more successfully than any other president.

In *Speaking Out: Inside the Reagan White House,* Reagan's former press secretary Larry Speakes expressed his view on the president's acting career: "Above all, he is an actor, and we never apologized for his Hollywood background. Communicating is a key ingredient of leadership, I always maintained, and if being an actor made him a better communicator, then so be it."

On both radio and television, Reagan was a polished speaker who roused public sentiment and goodwill. Whether or not people agreed with his policies, his style of communicating and his sincerity impressed them. Analysts believe that Reagan's appealing personal qualities, his expressions, and his ability to tell an appropriate story "played well" to television audiences. Reagan

also had a very talented speech writer named Peggy Noonan. Sometimes Reagan publicly misstated a fact or figure, but the media and the public were forgiving. His press secretary and staff usually cleared up the mistakes the following day with minimal damage to the presidential image.

With the help of aides, Reagan created TV news events that kept him in the public eye. When Reagan went to England for a meeting, his aides arranged for a side trip to Normandy, France. American troops had landed there to help break Germany's grip on Europe in 1944. Accompanied by old American soldiers who were in the landing parties, Reagan delivered a dramatic speech before TV cameras. At home, Americans tuned in to the nightly news and saw a moving speech made more memorable by the presence of the war veterans who risked their lives for freedom. "It turned out to be one of Reagan's best news events," wrote Speakes.

Does Television Affect Voter Turnout?

In 1960, television proved that it had the power to persuade voters. With a record 64 percent of voters casting their ballots in the Kennedy-Nixon race, televised debates had focused national attention on the candidates. In a survey conducted after the election, over half of those who voted said that these debates had influenced their choice of candidate. Just after Kennedy had won by a narrow margin of 112,000 votes, he remarked, "It was TV more than anything else that turned the tide." According to Philip Seib, "Many voters' ballot decisions are not based upon careful analysis of issues and candidates' background but on an instinctive feeling of whom they can trust. Television is a useful tool for this kind of decision-making."

In the presidential election of 1988, voter turnout set another record. This time, however, the number of voters dropped to the lowest level since 1924. Less than half of all eligible voters exercised the right to select their president. The issue of nonvoting is complex and different for each campaign. But during the presidential race of 1988, citizens complained that neither Republican candidate George Bush nor Democrat Michael Dukakis appealed to them. Bush was vice-president under Ronald Reagan from 1980 to 1988. Dukakis had been the governor of the state of Massachusetts. Rather than choose between the lesser of the two evils and feeling powerless in the electoral process, many voters stayed home on election day.

Negative television advertising may have caused further voter disenchantment with the candidates. Bush ads blamed Dukakis for the serious pollution problem in Boston Harbor. In this ad, images of dead fish, floating trash, and scum blanketed the TV screen. The ad that got the most attention, however, implied that Dukakis was "soft on criminals" and allowed convicted murderers out of jail on weekend passes. One such criminal, Willie Horton, had obtained a pass and committed serious crimes when he was released. While on leave, he terrorized a family, repeatedly raped a woman, and stabbed her fiancé. The ad omitted the fact that this program for criminals was set up not by Dukakis but by his Republican predecessor.

George Bush and Michael Dukakis shake hands before beginning their televised debate on September 25, 1988. Negative television advertising during their campaigns may have discouraged many citizens from going to the polls that November.

One survey conducted after the election "concluded that the assault on Dukakis had been responsible for Bush's victory." Wanting to avoid such a direct attack on his opponent, Dukakis resisted pressure from advisors to fight back with his own ads about Bush. "It might be argued I should have fired back immediately, but there would have been less voter participation," Dukakis said after the election.

Observer and writer Austin Ranney believes that television saturates voters with too many distasteful campaign advertisements. They grow tired of hearing candidates criticize each other and so withdraw their support for any candidate. When voters do not support any candidate, they usually do not vote at all.

Television networks also contribute to nonvoting in another sense. As the

Citizens exercise their right to vote for the candidate of their choice at the polls in a Texas precinct. The lowest voter turnout in over sixty years demonstrated voters' dislike of candidates' campaign tactics and personalities in the 1988 presidential election. Most people's experience of the candidates and their campaigns came from television.

votes are tallied for presidential elections, networks use complex formulas to predict the outcome. This process is called exit polling. The early projections are sometimes broadcast hours before polls have closed around the country. Because of the three-hour time difference between coasts, polls on the West Coast are still open after those on the East Coast have closed. Even before many West Coast voters have cast their ballots, networks have forecast the winners. West Coast voters then do not bother to go to the polls. In 1980, many voters were discouraged from going to the polls when President Jimmy Carter declared that opponent Ronald Reagan had beat him. When the announcement was made, polls on the West Coast were still open.

In the presidential election of 1988, CBS newsman Dan Rather was the first to name George Bush as the winner. At 9:17 P.M. eastern standard time he proclaimed, "It's over." ABC and NBC followed with their announcements soon afterward. The time was 6:17 P.M. on the West Coast. Many voters had not even made it to the polls yet. Though the networks disagree that they influ-

ence voters, Congress wants them to refrain from broadcasting state results until all polls around the country have closed.

Voters may have other reasons for not voting. They are sometimes skeptical about the information put out by the White House and the government. The roots of this disillusionment with government can be traced back to the

Former President Jimmy Carter announces on TV that the 1989 elections in Panama were fixed by crooked officials.

Senators Hubert Humphrey and Lyndon Johnson, both considered leading candidates for presidential nomination in 1960, pose together for photographers. Later, in 1964, Johnson ran for president with Humphrey as his running mate.

about the war. They lost faith in the president and in the government. In a way, television drove a wedge between the people and their leaders. This wedge still exists to some extent today.

Politics, Power, and Persuasion

Television has changed the way we elect officials, the type of people elected, and the way they conduct business after they take office. Politicians have to do more than present their ideas before the public—they have to acquire the right media image. Abraham Lincoln, often considered a great president, may not have been elected if he had had to campaign in the electronic age. He was tall, awkward, and had a squeaky voice. Under the close eye of television, these qualities may not have been judged suitable for the presidency.

When Walter Mondale lost the presidential election to Ronald Reagan in 1984, he recognized the dominant role of television in the campaign: "I think that, more than I was able to do, modern politics requires a mastery of television."

Success of political campaigns can be the result of an effective TV campaign. Television's power to affect the outcome of elections is a political fact of life. For the candidate who falters and fumbles before the TV cameras, he or she faces a much harder sell. Unfortunately, the one who shines the brightest on the TV screen may not be the most qualified to do the job.

late 1960s and early 1970s. Former President Lyndon Johnson was in office, and American soldiers were fighting in the Vietnam War. Government reports on the war conflicted with what the nightly news reported on TV. The message from the White House was that America was winning the war. But the message from the battlefields sent via TV was just the opposite. Almost every night, television news showed viewers the death and destruction in Vietnam.

Many viewers believed that Johnson and other politicians were lying to them

■ ■ ■ ■ ■ ■ ■ ■ ■ CHAPTER **7**

The Impact of Television

Television plays a big part in the lives of most people. Virtually every home in the United States has at least one TV set. Countless others have two or more. In 1980, the average set was on for about 6 1/2 hours a day. In 1989, that figure jumped to just over 7 hours, according to *TV Guide*.

People everywhere seem to be watching more television. About 2 1/2 billion people around the world watch TV. From Sri Lanka to the ghettos of Harlem in New York City, TV is an important part of many people's lives. TV exists in nearly 160 countries. Televi-

sion sets pop up in some unexpected locations as William Shatner, star of the TV show "Star Trek," discovered. Shatner noted that while he was filming near the Caspian Sea, a landlocked body of water between the Soviet Union and Iran, the following incident occurred:

> "We were in some deserted village in the wilds of the Caspian Sea. This waiter came up—of course he didn't speak any English—the man was as remote from civilization as you can get. He said, 'Capt. Kirk?' It was bizarre. In the back room of the restaurant where I was eating, on this ancient black-and-

Members of the cast of the 1960s hit TV series "Star Trek" have gained international fame because reruns of the program are shown worldwide. The series' characters are so familiar to people of the TV generation that teachers often use them in the classroom as examples of human character traits.

white television set, 'Star Trek' was playing."

Although television is popular worldwide, in the United States, watching television is a favorite pastime. Only two other activities take up more time—sleeping and working at a full-time job. For some, television plays the role of companion. They are soothed by its comforting presence. Young people spend more time in front of the TV set than they do in school or playing. By the time the average young viewer reaches the age of sixteen, he or she has watched about fifteen to twenty thousand hours of television.

Young people typically manage to squeeze in about twenty-five to thirty hours a week in front of the TV set. Often, other activities are going on at the same time. Children play with their toys, eat, wander in and out of the room, and do their homework.

With so much time spent watching TV, people have wondered and worried about its effects on viewers, especially children. Psychologists, sociologists, writers, and even Abigail Van Buren of "Dear Abby" have explored the pros and cons of TV. Experts remain divided on how TV viewing affects society, individuals, and educational achievement.

The Social Impact

Television informs and educates viewers about the world around them. The medium allows viewers to explore and better understand aspects of their own culture, and it acts as a bridge to other societies. TV has been described as a window to the world. Through this window, events that change the world can be observed, such as men walking on the moon, the unseating of communism in Eastern Europe, and the unprecedented student protests at Tienanmen Square in China. Without leaving the living room, the average person is transported to other countries, other cultures, and even beneath the sea.

Events and people come alive on TV and can move viewers to respond like no other medium. In 1985, for example, viewers were shocked into action by tele-

A moonwalking astronaut plants the United States flag on the lunar surface in 1969. This action was seen by millions of TV viewers back on earth. The medium of television now lets the average person witness events of historic significance—even if they happen on the moon.

baby-sitter, children are bound to be influenced by what they see. Writer Kate Moody believes that TV has taken over the job of raising children and that "the transmission of culture to children does not come as much through the family as it comes from outside—mainly via television."

Though the "electronic baby-sitter" is no substitute for parental supervision, numerous TV programs reflect positive family values and thoughtful plots. Family shows like "Growing Pains" and "The Cosby Show" offer good role models and sound solutions to social problems like drug use. Other shows, such as "Mr. Rogers' Neighborhood," make a special effort to teach self-esteem and cooperation to young children. If these shows are succeeding, then perhaps children are learning positive social values that they can apply to

A German demonstrator pounds the Berlin Wall with a sledgehammer as East German border guards look on. Television allowed us to share in the historic moment of triumph as the wall was torn down piece by piece by freedom-hungry East Berliners in 1989.

vision reports and pictures. For some time, newspapers and magazines had reported on the horrible famines in Africa. But Americans did not seem to notice until TV cameras brought the images of starving, diseased babies into their homes. Donations of cash, food, and supplies were made immediately, thanks to the visual impact supplied by TV.

Despite the benefits, some people worry about the negative side effects of television. They are scared that TV, instead of parents, is teaching social values and attitudes. Much of television is a reflection of American values, beliefs, and rules. When parents use TV as a

Chinese college students clasp hands in a show of solidarity during demonstrations for democracy in Beijing in 1989. The students knew that television would bring their cause into the homes of people worldwide.

TV programs like "Growing Pains" (top left) and "The Cosby Show" (top right) attempt to depict modern American family life in a positive light. Such programs can provide socially approved role models for viewers. The darker side of life, like the ruthless quest for power at the expense of others, is also addressed in more adult-oriented shows like "Dynasty" (cast shown at bottom left) and "Falcon Crest" (scene at bottom right). Programs like these can have a negative influence on immature viewers.

their lives. On the other hand, they may pick up negative values or actions from other shows.

Sometimes television contributes to social trends. The illusion of wealth and glamour thrives on prime time television. Shows like "Falcon Crest" and "Dynasty" portray fabulously rich people. The main characters live in large homes with servants, drive expensive cars, and wear beautiful clothes. Some observers say that this focus on wealth and possessions has helped create a trend toward materialism in society. From designer underwear to expensive cars, Americans are more concerned with earning money today than they were fifteen years ago. Americans want expensive products. They want to feel and look prosperous, even if they cannot really afford it.

The Psychological Impact

For years, people have wondered how TV affects personality, behavior, and emotions. Television has been criticized for irresponsible displays of sex and violence. People worry that immature or unstable viewers will imitate these irresponsible behaviors in real life. At the same time, the medium has been praised for exploring serious issues, such as child abuse, AIDS, and mental illness.

But TV shows touch each viewer in a slightly different way. For many, it provides an avenue of escape from the pressures of work or school. "TV helps me relax after work. I can forget about the problems at the office," reported one viewer.

Although the fantasy of TV may divert and entertain adults, younger audiences may become confused. Younger viewers have a harder time than adults do in drawing the line between make-believe characters on TV and reality. Does Cookie Monster exist? Can Superman actually fly? Educational experts advise parents to discuss and explain TV characters and their actions to younger children. This will help them to discern what is real and what is imaginary. "Children may understand that

Neighborhood children gather at the home of a friend to watch "Sesame Street," an award-winning children's educational program. Some experts believe that fantasy creatures, like Big Bird, when seen on TV, confuse children about what is real and what is make-believe.

what they see is made-up, but that does not stop them from believing it. It now appears programs that are obviously fantasy can be judged by children as at least partly real," wrote Neala Schwartzberg in *Parents* magazine.

TV can confuse a younger viewer's ideas about reality in another way. TV shows shorten the length of time it takes to solve problems. Within thirty to sixty minutes, TV families clear up misunderstandings, criminals are brought to justice, and diseases are cured by caring physicians. The characters, especially in family sitcoms, are wiser and happier after resolving their conflicts. In reality, many problems are never fully solved, and the process is rarely as smooth as television portrays it.

Because television has to make the most of each minute, relationships are often oversimplified. Television condenses the blossoming of new relationships and skips over important stages of growth. For example, on "The Love Boat," men and women from totally different backgrounds meet on glamorous cruises. Within just a few days, serious romances occur, and they often end in marriage. In real life, most relationships take more time to develop.

On the positive side, the companionship of TV may be helpful to children who come home to an empty house after school. These so-called "latchkey" children may spend several hours by themselves because both parents work. The president of Action for Children's Television, Peggy Charren, believes that TV may be important for them. "I really believe that for the latchkey child home alone after school, TV is one constant in the child's life. Without it, children would *really* be alone in the house. . . . Even watching

soap operas is better than being nervous about being alone."

Other groups of people also find emotional comfort in the presence of television. People who are physically restricted or handicapped may rely heavily on television, more so than the average person. Residents of nursing homes, people who are paralyzed, and the chronically ill may feel less isolated from the rest of society because of TV.

Television and Learning

When television was introduced to mainstream America in the late 1940s, educators were worried. Some believed that TV was a threat to the educational system and would destroy student motivation to learn. The president of Boston College, Dr. Daniel Marsh, summed up his frustrations in a speech before the 1950 graduating class: "If the television craze continues with the present level of programming, we are destined to have a nation of morons."

Heated debates still rage about the effects of TV on learning. The National Parent-Teacher Association (PTA) believes that "children who watch a lot of TV get lower grades, put less effort into their schoolwork, have poorer reading skills, play less well with other children and have fewer hobbies and outside activities than do children who watch an hour or less of TV per day."

A psychologist at the University of Massachusetts, Daniel Anderson, has done research that supports a different viewpoint. In a study with 165 children, Anderson found that TV had little effect on learning ability. Anderson concluded that television did not cause children to do poorly in school or cause them to spend less time playing. He found that homework done in front of the set was no better or worse than homework done under other conditions.

Anderson also discovered that TV did not create shorter attention spans, as some critics claim. "What surprised people about this study is that there is so little evidence to support common beliefs held about television," he said. The debate over TV and learning promises to be an enduring one. But educators and TV producers have found some common ground in the form of TV shows that teach as well as entertain.

One such show is "Sesame Street." The show attempts to teach the alphabet, counting, social values, and self-respect to children. On the air since 1969, one of its original goals was to help prepare preschoolers for a positive school experience. The writers and producers of the show were especially concerned about reaching out to poor and minority preschoolers. The show, however, at-

tracted children of all ages, from poor and rich families alike.

"Sesame Street" makes learning fun. The main character is Big Bird, a tall, yellow creature who reads books and tells stories. Through skits, animation, and songs, viewers learn about geography, the environment, and social skills. Real-life experiences are also part of the lively action. When one of the actresses on the show became pregnant, a story was done about the baby. Follow-up shows dealt with finding a day-care service for the baby so that her mother could return to work. Finding a good day-care facility is an important issue for many families.

Despite its loyal following, "Sesame Street" has not been overlooked by critics. Some claim that a child exposed to the fast-paced action of such shows will be bored when he or she has to adjust to the slow, steady pace in the classroom. One teacher described her frustrations. "I can't compete with television. I can't change my body into dif-

The "Sesame Street" cast includes puppets as well as humans. These characters have become household names recognized by almost every child today. The aim of "Sesame Street" is to make learning fun. But it perhaps makes learning too much fun. Classroom teachers complain that they cannot compete with the action-packed, colorful approach possible on TV, so their students tend to get bored in the classroom.

ferent letters, nor can I change color. The lessons I consider exciting fall flat because I don't do these phenomenal things." Other educators complain that the popularity of the show "promotes the TV habit at a young age."

Television and Reading

Today, television is children's primary waking activity—not playing, studying, or reading. For some children (and adults), reading at home has been replaced by TV. "In fact, about half of all fifth-grade students spend only four minutes a day reading at home, as opposed to 130 minutes a day watching TV," writes Dr. Melitta Cutright, author and communications director for the National PTA in Chicago.

Watching TV makes learning to read more difficult, according to some researchers. It conditions the mind for short segments, fast action, and quick cuts to other scenes. This "training" may hamper reading because this skill requires thought, focus, and longer attention spans. Thinking about what is being read activates the reader's brain cells and forces him or her to think. With the exception of a few instructional programs, television requires little mental work from the viewer.

Do TV fans turn on their sets and turn off their brains? Possibly, according to one scientist. Dr. Erik Pepper, an expert on brain waves, studied the brain activity of TV watchers. He found that brain waves changed dramatically after only a few moments of viewing. Most of his subjects showed an increase in alpha waves, which are associated with sleep and relaxation. Television may literally put viewers to sleep.

When used in conjunction with classroom instructions, some educators have found that TV encourages reading. In one experiment, a group of junior high students was asked to read a short story. Another group read the same story and saw it on film. The group that saw the story on film enjoyed it more, had better comprehension, and remembered it longer. Because the experience was so positive, the students actually wanted to read more stories. Children who found school to be difficult seemed to benefit the most by seeing the story on film as well as reading it. "Contrary to popular opinion, books and television need not be two media at war with each other," observed writer Patricia Marks Greenfield.

Reading and TV overlap in another way. For certain educational shows, television networks send schools advance copies of scripts to be read in the classroom. Millions of scripts with teachers' guides are delivered to schools each year. Students study the script and later watch it come alive on the TV show. In Philadelphia, Pennsylvania, teachers found that the scripts encouraged more interest in reading and that comprehension rose.

Television Violence and Aggressive Behavior

Ever since the public welcomed TV into its home in the late 1940s, television has been under scrutiny. Cop shows and westerns were blamed for the rise in juvenile delinquency and vandalism in the 1950s. "Gunsmoke" was among the first shows cited for too many shootings and fistfights. For decades, civic groups, the government, and parents have been concerned about the effects of TV violence on viewers, particularly the young.

Does violence on TV cause people to commit crimes? Are heavy TV watchers more aggressive than those who watch in moderation? Who should control what a child sees on TV—parents or producers? These are some of the basic questions that have plagued the television industry and led to government inquiries and psychological studies.

As early as 1954, the federal government entered the debate. Senator Estes Kefauver led a congressional investigation on juvenile delinquency. After hearing expert testimony from educators and psychologists, he concluded that TV contributed to juvenile crime. More hearings followed a decade later when Senator Thomas Dodd objected to the "television industry's unnecessary use of violence." In 1969, the government initiated a major investigation into TV violence. Armed with $8 million, the investigation was organized by the surgeon general, Wilbur H. Stewart. In 1971, he reached a cautious conclusion: evidence suggests the existence of a relationship between TV violence and aggressive behavior by viewers.

But many people disagree with these findings. One observer, Martin Maloney, thinks that we cannot blame all the ills of society on TV. He believes that it is silly to say that TV has such awful effects on everyone. "But if you really wonder why people beat their children, or take heroin, or bomb civilians, or pollute the air they themselves must breathe . . . well I hope you find a scapegoat [someone to blame] more convincing [than a television producer]."

Whether television causes people to behave aggressively is still an unanswered question. What is certain is that

if people did not watch violent shows, such shows could not affect them. Why, then, are people attracted to violent TV shows?

Writer Edith Efron notes that viewers are drawn to crime shows because of the heroes, not the villains. Viewers want to see the cop or the hero prevail. The "good guy," not the villain, keeps the viewers coming back for more.

The late Dr. Bruno Bettelheim, a child psychologist, offered another theory. He believed that the aggression viewed on TV helps children vent their own fears and frustrations. This is called the catharsis effect. Children's shows are like fairy tales and allow children to indirectly confront the scary people and things in their lives. The shows may help rid young viewers of tensions, according to Bettelheim.

Other psychologists have studied the effects of televised acts of violence

Senator Estes Kefauver delivers a speech at a convention. Kefauver led a congressional investigation of juvenile delinquency in 1954. He was convinced that TV contributed to juvenile crime.

Noted University of Chicago child psychologist Bruno Bettelheim speaks at a conference in New York City. Bettelheim believed that TV helps children deal with their fears and aggressive feelings by giving them a "safe" outlet for such feelings.

and reached different conclusions. After watching shows with aggressive acts in them, children tended to behave more aggressively afterward. Arguing, pushing, and even hitting were ways that aggressive behavior surfaced. Some of the aggressive acts seen on the shows were repeated months later.

Reports from other behavioral labs also point an accusing finger at TV. According to some psychologists, young viewers who watch violence on TV seem more willing to inflict pain on others. When a problem arises, they try to resolve it by arguing instead of talking. Research has also shown that serious TV watchers are less sympathetic to the pain and hardships that others experience. Because they have witnessed so much pain, injury, and death on TV, they may become insensitive to it in real life.

Changing Trends

During prime time dramas, up to five violent acts may appear in an hour, according to one research study. Yet the overall trend in prime time violence may be on the decline. One study noted a dramatic reduction of violence from 1985 to 1987.

Though evidence suggests less aggression in TV for adults, Saturday morning cartoons are a different story. Action and adventure cartoons easily depict brutal acts, such as clubbing a character with a baseball bat or throwing a plunger at a cat's face. This action is what holds the young viewer's attention, rather than the plot or the cute characters. Of course, cartoon characters always bounce back and recover from these acts that would kill or maim if done in real life. Since younger viewers have trouble deciding what is real and what is not, they are more prone to copy the actions of cartoon characters. They may not realize the consequences.

TV on Trial

In one case, television violence inspired a real-life assault. A show called "Born Innocent" fueled the imagination of a teenage gang of girls. "Born Innocent" was the fictional account of a teenage girl who was victimized by her family and the child justice system. Raised by an alcoholic mother and an abusive father, the girl coped with her family situation by running away—again and again. Unable to control her, the girl's parents confined her to a detention home with drug addicts and child prostitutes. Instead of getting the emotional help and understanding she needed there, the main character was treated cruelly by the detention authorities and other female inmates. The story reached a terrifying climax when female inmates sexually assaulted the girl with a broom handle while she was taking a shower.

NBC aired "Born Innocent" in 1974 with little fanfare at first. Several days later, the story was replayed for real on a deserted beach near San Francisco. A gang of girls decided to reenact the violent rape scene by attacking two young girls in a similar fashion. The parents of one victim took NBC to court and claimed that the violent episode of "Born Innocent" was responsible for the assault on their daughter. After four years of legal battles, NBC finally won its court case and was relieved of any responsibility for the show.

Though the victim's family lost the court case, critics of TV violence had made their point. This case, they maintained, was "proof of the harmful effects of unrestrained television on impressionable minds."

Outcries of censorship are often heard from the networks when they are criticized. In a free society, their rights of self-expression should not be denied, the networks argue. The parents, not the networks, should monitor what children view. While millions of viewers see acts of TV violence every day, few imitate them. Networks have no control over the unstable individuals who are driven "over the edge" by the content of a show.

Putting TV in Perspective

Audiences have given TV mixed reviews since its beginning. But now more than at any time before, audiences have

Fred Rogers, star of the popular children's show "Mr. Rogers' Neighborhood," flashes his winning smile at the camera. Rogers believes that television is a marvelous and beneficial way to educate, stimulate imagination, and teach problem-solving. Watching TV can be emotionally and educationally rewarding, Rogers claims.

many more programming choices. When TV fails to live up to its potential, viewers need not sit by and accept programming that offends or seems meaningless. They too have the right to self-expression. They can write letters to the networks, the FCC, ACT, and even the show's sponsors. Simply changing channels or turning the set off are options available to viewers of any age.

On the bright side, selective TV viewing can be emotionally and educationally rewarding. Fred Rogers' of "Mr. Rogers' Neighborhood," summed up the benefits of television in a speech he made: "It is a marvelously wonderful thing that can be used to nurture, to stimulate imagination, to disseminate information, to show how problems can be worked on and often solved by mutual respect. . . . And to demonstrate the diversity of human beings."

Television of Tomorrow

As the next century approaches, television is expected to maintain its lead as the most popular form of mass communication and entertainment. For most Americans, TV is expected to dominate their free time and continue to successfully compete with the motion picture industry for viewers. Improvements in the quality of the TV picture and sound will rival that of motion pictures, thus strengthening TV's hold on audiences.

Cable television will continue to be a popular option. Cable TV was originally designed to send TV signals to remote areas of the country. Regular TV signals, which traveled through the air, were unable to reach some places, so a special underground cable was used to send the signals instead. Cable TV also differs from regular TV because customers must pay to receive it. Even in urban areas, people pay for this service because it offers a large variety of alternative programming. In the early 1980s, the government eliminated many of the rules governing the television industry. This action was called

TV news anchors announce the latest news stories on the Cable News Network. The award-winning CNN has broadcast news round-the-clock on cable TV since 1980. Cable television has enriched TV viewing considerably by providing a larger menu for viewers to choose from.

deregulation and made it much easier for cable companies to do business. Since then, cable has spread to over half the homes in the United States. Cable TV is perhaps the most important competitor of the major commercial networks—ABC, NBC, and CBS.

One way that TV in general will retain its popularity in the future is by appealing to America's different cultural groups. Already some networks specialize in broadcasting shows in Spanish. They have captured large segments of the Hispanic audiences of Los Angeles and Miami. Makers of products and advertisers have recognized the purchasing power of the Spanish-speaking population and want to cultivate this growing market. Procter & Gamble, PepsiCo, and other companies plan to spend millions on commercials made in Spanish. Kelley Gillespie, a representative of Procter & Gamble, said, "Hispanic media give us a good opportunity to acquire a large group of loyal consumers."

The decade of the 1980s saw the major networks—ABC, NBC, and CBS —loosen their grip on the public airwaves. Serious competition emerged from the Public Broadcasting Service (PBS), Home Box Office (HBO), and other new networks. With an explosion of more programming choices, many viewers abandoned the big networks. Looking forward, competitors will continue to entice more viewers away from big network channels. As audiences dwindle, major networks will respond with creative strategies to lure them back.

Europe is also planning some changes in the TV industry. For a long time, the United States has sold its programs to other countries. Up to 70 percent of all shows broadcast throughout Europe originate in the United States.

"Miami Vice" stars Don Johnson and Philip Michael Thomas, and series guest star Glenn Frey pose menacingly in this scene from the TV crime drama. "Vice" and other popular TV shows have been exported to foreign countries where they have also proven immensely popular.

"Dallas," "I Love Lucy," "Miami Vice," and even "Sesame Street" have entertained audiences abroad for years. They may have even overwhelmed European culture to a degree. TV shows have not only exported an American brand of entertainment but also American values, trends, and life-styles. Some nations worry that American culture will eventually replace their own rich heritages.

The minister of French culture, Alain Le Diberder, explained how American TV affects his country. "Our views of the world and our consciousness and opinions are organized and provided by TV. When we imagine the rest of the world, we don't think of France or Europe. We imagine America."

Europeans want their airwaves back. The European Common Market (ECM), an organization concerned with the economic state of Europe, has led the fight. In 1989, the ECM directed that at least 50 percent of airtime be slotted with European-made shows. Jacques Delors, president of the ECM, justified the action. He said if changes are not made, "in fifteen years, all of our TV sets will be Japanese, all of our programs American, and all of our viewers European."

Since the ECM directive does not carry the same weight as legislation, whether or not European networks will comply with it is uncertain. What is certain, however, is that American TV executives will fight to preserve a very lucrative market for their products. As the battle heats up, Congress may step in to protect television interests abroad. American networks have already pressured Congress into considering trade restrictions on European goods as a way to convince the ECM to change its stand.

Though some foreign nations want to halt the spread of American television, others do not. The tiny island of Rarotonga in the South Pacific recently ventured into the age of television. The largest of the Cook Islands, three thousand miles south of Hawaii, Rarotonga's first TV station began operating on Christmas Day in 1989. Many of the programs will be American. Some islanders fear that TV's influence will alter their quiet culture and traditional ways. "We'll probably have to drag kids away from the TV set to get them to plant crops. It will make people lazy," said Jillian Sobieski, a resident of nine years.

The government official in charge of broadcasting on Rarotonga is Tiki Matapo, and he disagrees. He thinks his people will benefit from television. As TV spreads to the other Cook Islands

The whole cast of "Sesame Street," both human and puppet, poses for this portrait on the show's set. "Sesame Street" has entertained audiences in dozens of countries around the world for years. Some foreign leaders worry that American culture, as seen on TV, will overwhelm their own native cultures.

in about three years, Matapo predicts that it will unify and educate the islanders.

The spread of television to remote corners of the world is made possible by satellites. Since it is impractical to station transmitter towers all around the globe, carrier waves are instead beamed to a satellite orbiting the earth. Carrier waves are sent from a ground transmitting station on earth up to the satellite in space. The satellite amplifies and reroutes the waves to ground receiving stations far from where the waves started, often on another continent. Orbiting more than twenty-two

LIVE BROADCASTS AROUND THE WORLD

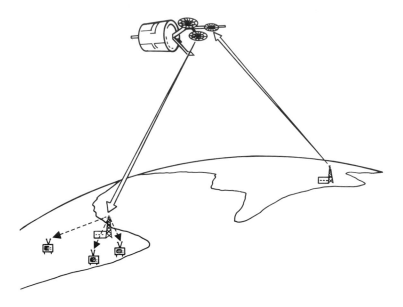

Communications satellites orbiting 22,500 miles above the earth can relay television signals from one continent to another. The broadcast is transmitted on radio waves from a radio tower on the ground up to the satellite. The satellite receives the audio and visual signals, amplifies them, and sends them on its own radio waves to powerful ground antennae located at transmitting stations. The transmitting stations then broadcast the audio and visual signals to individual television sets.

thousand miles up, the satellite is powered by the sun's rays. Thousands of solar cells cover the satellite's exterior like shingles on a roof. The solar cells are designed to turn the sun's energy into electricity for power. Power is stored in batteries in the upper section of the vehicle. Satellites are also used to relay telephone calls.

Isaac Asimov, writer and futurist, believes that the marriage of TV and telephone communications satellites will one day produce a planet that is one cultural unit. With a single, worldwide culture will come a common understanding supported by shared values and objectives. Perhaps then people will have the determination to unite against global problems, like hunger and disease. "And maybe then we will survive," said Asimov.

High-Definition TV

By the end of the 1990s, razor-sharp images will replace the TV pictures of today. A new invention called high-definition TV (HDTV) holds the key to bringing the world into sharper focus. Not since 1945 when the image orthicon camera was invented has a major change in picture quality occurred. Though HDTV is considered highly ad-

Science-fiction writer and futurist Isaac Asimov foresees a new unity among the nations of the world made possible by global communications networks. These networks will consist of combined television and telephone technology.

vanced technology, it is based on a fundamental principle of the earliest experiments with TV technology— controlling the number of scanning lines. Still in its experimental stage, HDTV's clarity will be achieved by increasing the number of scanning lines from 525 to perhaps as many as 1250. Electrons will carry much more information about the images to the TV screen. As electrons scan the back of the screen, this added information will be displayed by building an image with many more lines and thus greater clarity.

HDTV faces not only technical hurdles but obstacles of government regulation. HDTV must comply with the FCC standards. Just as the FCC set rules and standards for commercial and color television in the forties and fifties, it will do the same for HDTV. The FCC has already ruled that HDTV systems must be compatible with existing televisions and not render them obsolete. More FCC standards governing HDTV lie ahead, but there are still many unknowns that trouble the television industry.

Some American companies are gambling on the future of HDTV, despite the uncertainties. The Zenith Corporation has invested vast amounts of time, talent, and money into HDTV research. North American Philips plans to build a $100 million manufacturing plant to produce HDTV picture tubes as well as conventional tubes. Iva Wilson, president of the display division at Philips, said, "We have to take the risk in order to ensure the future of HDTV."

Japan, Europe, and the United States are racing to perfect HDTV. The winners stand to earn a fortune from the sales of the new TV sets. For over twenty years, Japanese scientists have studied and experimented with equipment to produce clearer images. They have spent more than $1 billion. As of 1990, they seem closest to producing the first commercial high-definition TVs. European companies developed a serious interest in HDTV in the early 1970s and have pumped about $200 million into research. As the race nears its finish, consumers can look forward to better viewing. Though no one can say for certain, HDTV is expected to make its way to store shelves between 1995 and 2000.

Communication through Interactive Television

In the past, television was a one-way communication device. Electrical signals traveled to homes with no re-

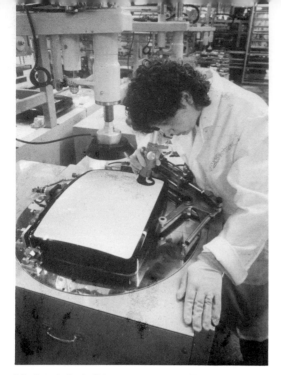

Mechanic Judy Bisso uses a specially designed microscope to check optical alignments in a television picture tube at a Toshiba Westinghouse Electronics Corporation factory in Elmira, New York.

sponse from the audience. Interactive television promises to make TV viewing a two-way process. Instead of sitting quietly in front of a TV set, a viewer will be

able to react to what is happening on the screen by simply pressing a button. TV sets would come equipped with a panel of buttons similar to a remote-control device that transmits audience reactions back to the television station.

The basic idea of interactive TV has been around since the 1950s. The public was first introduced to the concept by Winky Dink, a cartoon character. When he appeared on TV, children were invited to interact by outlining letters of the alphabet on a plastic overlay they placed on the screen, thus revealing secret messages. In the 1980s, newscasters sometimes ask audiences to participate in another way. When a controversial story airs, viewers are invited to call a certain number if they agree with a certain point of view and another number if they disagree. The calls are tallied by a computer at the TV stations, and the results broadcast later.

Some sports enthusiasts are already enjoying interactive TV on a limited basis. In New York City, viewers can subscribe to a service that allows them to

A Nielsen Media Research employee holds a remote control device as she demonstrates the new "peoplemeter" used to measure network television audiences. The peoplemeter (on top of the TV) automatically records which shows are being watched and by whom. Such information generates the "Nielsen ratings" by which sponsors and networks can judge the popularity of a program.

have control over what is seen on the screen. With the punch of a button, they can change camera angles during a ball game or select different levels of aerobic exercises. For a small monthly fee, viewers are provided with a remote-control device that signals an electronic unit on top of the TV set. This unit sends messages to a computer at the local TV station. The messages contain directions for the changes that viewers want to appear on their screens.

In California, a new interactive TV system makes baseball come alive. As the batter steps to the mound, viewers predict the outcome of the play. Their predictions enter a special computer terminal and are relayed to a computer at the TV station that keeps score. To make watching the game on TV more exciting, stations offer prizes to those who consistently make the correct choices.

In the future, interactive TV may play a big role in assessing public opinion. Viewers may be able to respond immediately to TV shows, registering complaints or praise. Advertisers could evaluate the impact of their commercials by obtaining immediate viewer reaction. When a politician makes a speech, public approval or disapproval could be quickly measured. Interactive television may someday allow citizens to vote for local and national candidates from the privacy of their homes. For physically disabled voters, this would be a great service.

Interactive TV may also have a place in the classrooms of tomorrow. In addition to the traditional approach to education, lectures or lab work could be televised. Special effects and graphics could be used to liven up lessons or illustrate complex ideas and concepts.

New York Mets pitcher Ron Darling gets ready to deliver his pitch. A California TV system now in operation lets viewers interact with the program by predicting how each batter will do as he steps up to the plate. Viewers enter their predictions into a special computer that relays it to the TV station.

More research and study is needed, however, before TV obtains its teaching credentials.

Television as a Family Helper

At home, kitchens of the future may be equipped with special TVs that help with meal planning. Sets will display a wide variety of tasty dishes on the screen. Cooks will browse through a vast collection of recipes, from pot roasts to strawberry shortcake, and see the prepared dish on the screen. In an increasingly health-conscious world, nutritional data will also be displayed. After the cook makes a selection, the TV will print out a recipe for the desired dish. Television of the future may

A Tokyo woman uses a Lumanaphone TV-phone to talk to a friend. A still photograph of her friend is projected onto a black-and-white TV screen. The Mitsubishi Electronics Corporation introduced this item to Japanese consumers in the fall of 1987.

help solve the timeless problem of what to fix for dinner. Besides helping out in the kitchen, TV of the future will assist in other areas of family life.

The telephone and television set may combine their capabilities in households of the future. When someone telephones, the caller's image will be displayed on the TV screen for a more personal conversation. Grandparents could watch their grandchildren as they speak into the phone. Teenagers might feel closer to their friends if they could see them during phone conversations. This type of system is already used in some businesses and is called PictureTel. The face-to-face contact makes business calls more personal and saves money on travel. Executives do not have to schedule meetings and travel to conferences as often as a result of the new technology.

A visual message center for homes and businesses is another possible use for TV of the future. Linked to the telephone, callers could leave messages by dialing a predetermined code. Each code corresponds to a different message that is printed on the TV screen. To read the messages, users would turn on the

set and select a special channel with all their phone messages written out.

In addition to serving as a message center, television and the telephone may team up for yet another joint venture. Using a system called Viewdata, users can dial a special telephone number and tap into a reservoir of data and

Brian Elliott, a 28-year-old design student in Pasadena, California, shows his prize-winning entry in Sony's "Television of the Future" contest.

Astronauts Neil Armstrong and Edwin Aldrin plant the American flag on the moon in 1969. This was one of humanity's shining moments that we were all able to share through the miracle of television. Television has changed our lives forever by bringing history right into our living rooms.

information that will appear on their television screens. Telephone lines will transfer the request to a central data bank, much like an electronic encyclopedia. The computer will speedily sort through its files and send the requested data back over the same phone lines. Viewdata could be used at home to help students with ideas for science projects or history reports. In the office, Viewdata would provide the latest financial updates or help news reporters do background research for stories.

Streamlining the shape of TV sets has always been a goal of electronics wizards. The first TV sets introduced at the World's Fair in 1939 were bulky and awkward compared to the sleek designs of this decade. If Japanese researchers have it their way, the sets of today may be considered old-fashioned within a few years. Shortening the picture tube is the key to making a more compact TV set. Researchers are experimenting with new technology to replace the long tube with a shorter, more compact light-emitting diode (LED). If they are successful, TV sets will become so thin that they can be hung on the wall like paintings. In many countries where homes are not spacious, thinner TV sets would be a great convenience.

A Final Word

The magic of television promises to linger, inspiring and entertaining generations to come. Why has the world embraced this black box with such passion? TV is the ultimate dream machine. It delivers an endless parade of live drama, make-believe romances, and news from around our world. Without leaving the living room, viewers are transported to places they could never afford to visit. TV introduces viewers to people of all backgrounds. Though TV sets are made of metal, glass, and plastic, the images and sounds they produce are emotionally powerful and touch each viewer in a special way. Through no other medium can an audience feel the agony of a dying president, the joy of birth, or the wonderment of men walking on the moon.

Glossary

■ ■

amplifier: Equipment that strengthens electrical signals. Amplifiers are often located at transmitting stations, on top of transmitter masts, in satellites, and inside TV sets.

antenna: A wire-frame structure that receives carrier waves composed of electrical signals. Signals travel through the antenna wire into TV sets.

broadcast: To transmit sound and/or images via electrical impulses.

carrier wave: An invisible airwave made up of electrical impulses. TV shows travel on carrier waves.

catharsis effect: The psychological process of relieving pent-up emotions without directly expressing them.

control room: The heart of the television studio. The control room is enclosed with glass and contains all of the technical equipment needed to produce a show and transmit it to the audience.

director: The person who generally has creative control over a TV show. The director is responsible for lighting, sound, camera shots, and the final "look" of the show.

electron: The smallest particle of electricity that exists inside an atom. Inside of the picture tube, electrons scan the image behind the screen.

electronics: The science of electrons and how they behave.

frame: A still picture that appears on a roll of film or on videotape. Frames are transmitted so quickly that the human eye sees rapid movement instead of individual pictures.

gaffer: A technician who works with lighting in the production of a TV show.

grip: The person who moves scenery around to create different scenes for a TV show in production.

iconoscope: Vladmir Zworykin's term for the camera tube. The camera tube converts the image into electrical signals.

image dissector: Philo Farnsworth's electronic scanning device that converted the image into electrical impulses.

image orthicon: A special TV camera that required less light than previous cameras when it was invented in 1945 by Vladimir Zworykin. It was a major advance in camera technology.

interactive television: The process of viewers reacting and responding to TV programs. A remote-control device is often used to allow viewers to press a button to register a response.

kinescope: Vladimir Zworykin's term for the picture tube or receiver.

network: A chain of many TV stations working together.

pan: The act of moving a camera back and forth horizontally.

producer: The person who has total financial and creative responsibility for a TV show.

puffing: In advertising, the accepted practice of extravagantly praising a product, often making it seem better than it really is.

run-down sheet: The outline used to guide shows for which there is no complete script.

satellite: A vehicle launched into space that rotates around the earth or another planet. Satellites can be used to relay signals for radio, television, and telephone communications to all parts of the world.

scanning: The process of "reading" a TV picture line by line using electrical impulses.

script: The written format of dialogue and actions that actors follow. The script is the backbone of a television production.

switcher: A large electronic device operated by the technical director of a television show. Camera shots appear on the monitor screens of the switcher, allowing the director and technical director to decide which shots are to be aired or used for filming.

TelePrompTer: A device that displays a script or speech that is read by a performer speaker, or newscaster. The text appears on a scroll that rolls by, thus displaying what should be read next. Using a TelePrompTer makes the delivery smoother because the speaker can look straight ahead or down at his or her notes.

transmitter: A device that sends carrier waves through the air as they travel to TV antennae.

vacuum tube: A sealed glass container with no air in it. Inside the container, electrons are manipulated to perform various tasks. The TV picture tube is a vacuum tube.

videotape: One method of recording TV shows. Videotape is a magnetic strip that can be instantly replayed.

For Further Reading

Sally Berke, *When TV Began: The First TV Shows.* New York: Contemporary Perspectives Inc., 1978.

James Burke, *Connections.* Boston: Little, Brown, 1978.

Christopher Griffin-Beale and Robyn Gee, *TV and Video.* London: Usborne Publishing Ltd., 1982.

Karen Jacobson, *Television.* Chicago: Children's Press, 1982.

Eurfron Gwynne Jones, *Television Magic.* New York: Viking, 1978.

David Macaulay, *The Way Things Work.* Boston: Houghton Mifflin, 1988.

Jeanne Oslon, "Philo Farnsworth: Forgotten Inventor," *Cobblestone,* October 1989.

Edward Stoddard, *Television.* New York: Franklin Watts, 1970.

Works Consulted

Howard Blumenthal, *Television: Producing and Directing.* New York: Barnes & Noble, 1987.

Les Brown, *The New York Times Encyclopedia of Television.* New York: Times Books, 1977.

Harry Castleman and Walter Podrazik, *Watching TV.* New York: McGraw-Hill, 1982.

Donald Clark, ed., *The Encyclopedia of Inventions.* New York: Galahad Books, 1977.

Donald Clark, ed., *How It Works Encyclopedia of Great Inventors & Discoveries.* London: Marshall Cavendish Books Ltd., 1978.

Peter Conrad, *Television: The Medium and Its Manners.* Boston: Routledge & Kegan Paul, 1982.

Donna Woolfolk Cross, *Media Speak: How Television Makes Up Your Mind.* New York: Coward-McCann, 1983.

Melitta Cutright, *The National PTA Talks to Parents.* New York: Doubleday, 1989.

Evelyn Kane, *The ACT Guide to Children's Television.* Boston: Beacon Press, 1979.

League of Women Voters Education Fund, *Choosing the President.* New York: Schocken Books, 1984.

Rick Mitz, *The Great TV Sitcom Book.* New York: Perigee Books, 1983.

Kate Moody, *Growing Up on Television.* New York: Times Books, 1980.

Edward Palmer, *Children in the Cradle of Television.* Lexington, MA: Lexington Books, 1987.

Lee Polf and Edna LeShan, *The Incredible Television Machine.* New York: MacMillan, 1977.

Jeff Rovin, *The Great Television Series.* London: A. S. Barnes and Company, 1977.

Philip Seib, *Who's in Charge? How the Media Shape News and Politicians.* Dallas: Taylor Publishing Company, 1987.

Larry Speakes with Robert Pack, *Speaking Out: Inside the Reagan White House.* New York: Scribner's, 1988.

Theodore White, *The Making of the President 1960.* New York: Atheneum, 1988.

Michael Winship, *Television.* New York: Random House, 1988.

Alan Wurtzel, *Television Production.* New York: McGraw-Hill, 1983.

Index

About the Author

■ ■

Lila Gano is a free-lance writer who lives on a forty-two foot powerboat near San Diego, California. Interested in various types of writing, she divides her time between preparing magazine articles, writing children's books, and editing a newsletter for a career development organization.

Ms. Gano is a member of the Society of Children's Book Writers and the American Film Institute. She has a master's degree in psychology and counseling.

Picture Credits

■ ■

Cover photo by Barth Falkenberg
Gene Anthony/Light Images, 54
AP/Wide World Photos, 12, 13, 15 (both), 19 (top), 28, 30 (both), 46, 48 (both), 56, 57, 58, 59, 60, 61 (top), 63, 64, 68 (top), 69, 70, 76, 78, 79, 80, 81, 82, 83, 86 (both), 87, 88 (both)
courtesy of Bulova Watch Company, 50
Leo Burnek Company, Inc., 53 (bottom)
Howard Frank/Personality Photos, Inc., 16, 23, 24, 31, 35 (both), 36, 37, 38, 39, 41, 43, 45 (top), 73 (bottom left)
The Hertz Corporation, 52
Library of Congress, 18 (both), 19 (bottom), 34, 45 (bottom)

Light Images, 74
Diana Lyn/Shooting Star, 51 (top)
National Aeronautics and Space Administration, 71, 89
courtesy of Nike, Inc., 51 (bottom)
Reuters/Bettmann Newsphotos, 67, 72 (both)
Fred Sabine/Shooting Star, 73 (bottom right)
M. Sennet/Shooting Star, 53 (top)
Shooting Star, 73 (top right)
Smithsonian Institution, 14
Trindl/Shooting Star, 73 (top left)
UPI/Bettmann Newsphotos, 61 (bottom), 62, 65, 66, 68 (bottom), 85